MICROBIAL ECOLOGY

MICROBIAL ECOLOGY

A Study of River Ganga

By

Dr. D.R. Khanna
Dr. Tarun Chugh

DISCOVERY PUBLISHING HOUSE
NEW DELHI

First Published - 2004

Reprinted - 2018

ISBN: 978-81-7141-850-3

Microbial Ecology
A Study of River Ganga

Published by:

DISCOVERY PUBLISHING HOUSE PVT. LTD.
4383/4B, Ansari Road, Darya Ganj
New Delhi-110 002 (India)
Phone: +91-11-23279245, 43596064-65
Fax: +91-11-23253475
E-mail: discoverypublishinghouse@gmail.com
sales@discoverypublishinggroup.com
web: www.discoverypublishinggroup.com

Printed at:
Infinity Imaging Systems
Delhi

Preface

The present book has been written on the facts studied, to reflect various aspects of different microbial ecology of river Ganga at Hardwar. The coverage also includes description of the abiotic and microbial parameters of water. The seasonal fluctuations of various physico-chemical parameters have been discussed in relation to different type of bacteria observed at this place. Besides these the authors has also reported the zooplankton and phytoplankton of the river Ganga and its relation with abiotic factors.

The book should form a very useful contribution for Limnologists, Biologists, Microbiologists and post graduate studies all courses in microbial ecology and limnology.

Authors

Contents

1

Introduction

A. GENERAL

Named for the Hindu goddess Ganga and regarded by Hindus as the most sacred river in the world and with whom people are sentimentally attached since the time immemorial. The Ganga flows about 2,506 kilometres in India (Fig. 1.1 and 1.2). The banks of the river host many important religious ceremonies, especially at Varanasi, Hardwar, and Allahabad. Hindu pilgrims travel to the Ganga to bath in its waters, hoping either to benefit from its supposed restorative properties or to die and be transported to Paradise. The Ganga's headstreams—the Bhagirathi, Alaknanda, Mandakini, Nandakini, Bhilangana, Dhauliganga, and Pindar—all rise in northernmost Garhwal region of Uttranchal state in India. The river is sacred to Hindus.

The Ganga Basin, one of the most fertile regions of the world and also one of the most densely populated, lies between the Himalayas and the Vindhya Range, embracing an area of more than 1 million square kilometres. The river rises as the Bhagirathi in a snowfield situated among three towering peaks in the southern Himalayas, issuing from an ice cave 3,139 metres above sea level. At the village of Devprayag 214 kilometres from the source, the Bhagirathi joins the Alaknanda to form the Ganga (Fig. 1.3).

After descending at 2,827 metres, at Hardwar, the Ganga cuts across the Shivalik hills and for the first time enters the great plain of India (Fig 1.2). Therefore, Hardwar is known as

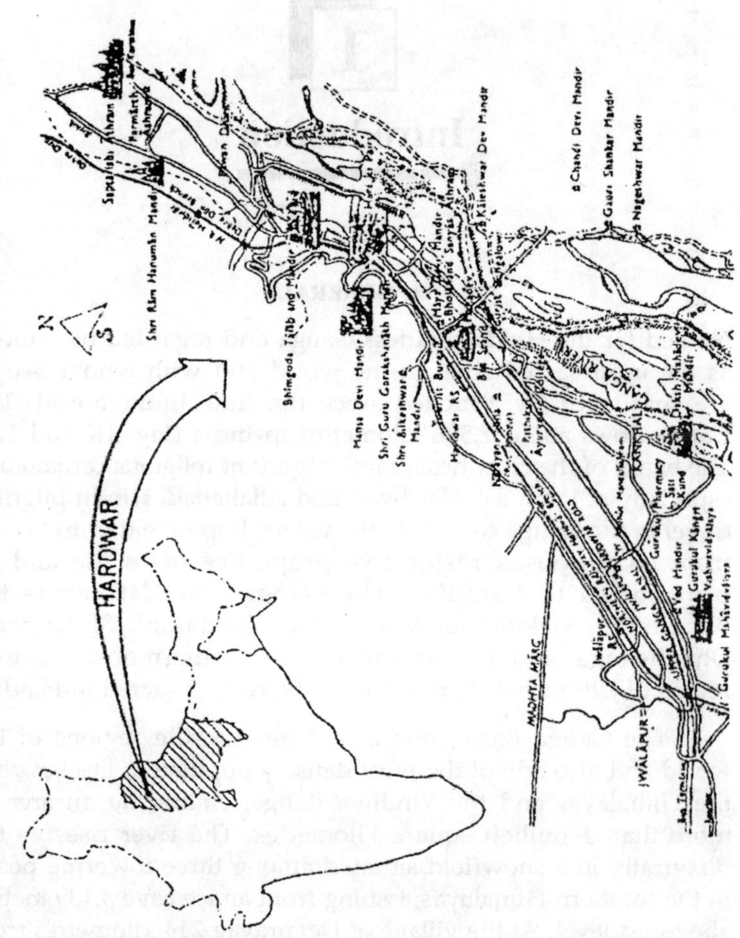

Fig. 1.1 Map Showing the Situation of Hardwar (Uttaranchal), India.

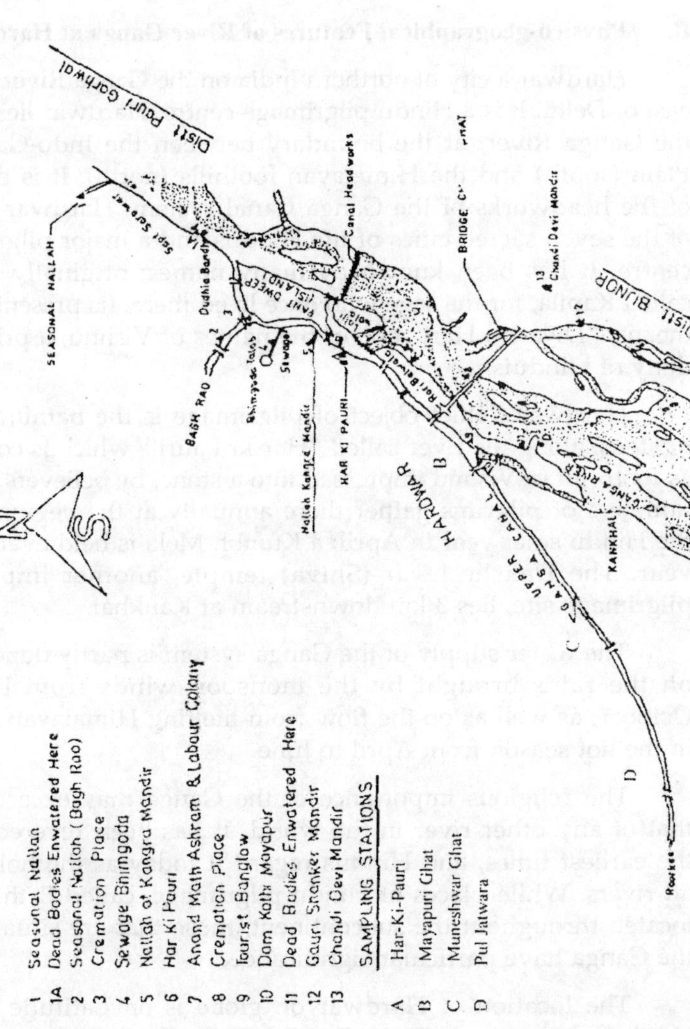

Fig. 1.2 Map of the River Ganga Showing Sampling Stations and other Important Places.

the "Gateway to God." From Hardwar it continues south and then south-east to Allahabad after a winding course of 785 kilometres, made un-navigable by shoals and rapids.

B. Physico-geographical Features of River Ganga at Hardwar

Hardwar a city of northern India on the Ganga River north-east of Delhi. It is a Hindu pilgrimage centre. Hardwar lies along the Ganga River, at the boundary between the Indo-Gangetic Plain (south) and the Himalayan foothills (north). It is the site of the headworks of the Ganga Canal system. Hardwar is one of the seven sacred cities of the Hindus and a major pilgrimage centre. It has been known by many names; originally it was called Kapila, for the sage who once lived there. Its present name means, "Dwar to Hari" one of the names of Vishnu, a principal deity of Hinduism.

Hardwar's chief object of pilgrimage is the bathing ghat, or steps, along the river called "Har-ki-Pauri", which is consider as footprint of Vishnu impressed into a stone, by believers. Large numbers of pilgrims gather there annually at the beginning of the Hindu solar year in April; a Kumbh Mela is held every 12th year. The Daksheshwar (Shiva) temple, another important pilgrimage site, lies 3 km downstream at Kankhal.

The water supply of the Ganga system is partly dependent on the rains brought by the monsoon winds from July to October, as well as on the flow from melting Himalayan snows in the hot season from April to June.

The religious importance of the Ganga may exceed than that of any other river in the world. It has been revered from the earliest times, and Hindus regard it today as the holiest of all rivers. While places of Hindu pilgrimage, called tirthas, are located throughout the subcontinent, those that are situated on the Ganga have particular significance.

The location of Hardwar on globe is on Latitude 29°58′ North and Longitude 78°10′ East while the height from sea level is 285.56 metres (Table 1.1). Present study was carried out on stretch of 8 km from Har-ki-Pauri to Pul Jatwara, which is having a width of 15-40 m and depth of 2-10 m. Total surface area and volume of water for this area was recorded 220000

Table 1.1: Physico-geographical Features of Hardwar

S. No.		Parameters
1.	Height (from sea level)	285.56 metre
2.	Latitude	29°58′ North
3.	Longitude	78°10′ East
4.	Length of River	8 km
5.	Width of River	15-40 metre
6.	Minimum depth	2 metre
7.	Maximum depth	10 metre
8.	Mean depth	8 metre
9.	Surface area	220000 metre^2
10.	Water volume	1760000 metre^3
11.	Ambient temperature in winter (°c)	9.3-16.5°c
12.	Ambient temperature in summer (°c)	26.8-40.4°c
13.	Ambient temperature in monsoon (°c)	12.9-28.5°c
14.	Climate	Temperate climate, Dry winter, warm/hot summer
15.	Annual precipitation	over 12.2 inches
16.	Precipitation in winter	2.1-3.2 inches
17.	Precipitation in monsoon	5.2-51.4 inches

metre^2 and 1760000 metre^3. Climate for this reason is temperate, dry winter and warm to hot summer with an ambient temperature for winter, summer and monsoon—9.3 to 16.5°c, 26.8 to 40.4°c and 12.9 to 28.5°c respectively. Annual precipitation here is over 12.2 inches (winter 2.1-3.2 and monsoon 5.2-51.4 inches).

Water being the most vital resource for all kinds of life on this planet. It is being adversely affected, qualitatively and quantitatively by all kind of human activities on land, in air and/or in water. Domestic sewage consists of about 99.99% water coming from washing, rinsing, flushing and other activities. The other materials are such as fecal material, cellulose, paper fibers, cloth, garbage and dissolved material (about 0.1%).

During recent years, due to increased population and industrial growth, the quality of Ganga water has deteriorated considerably from domestic and industrial sewage, which contains large number of chemicals and heavy metals. The waste materials react with each other, as a result, the water is polluted and may become toxic, which ultimately make water un-potable and also severely affect the bio-productivity of the aquatic system. At place like Hardwar, thousand of pilgrims take bath daily in the Ganga and degrade its water quality not only by bathing, but also by dumping the things like flowers, ash and bones of dead persons. Such activities are increased many fold on occasions like Kumbh, etc.

River Ganga is a major source of water for the residents in district Hardwar, but because of its location near major population centres it is heavily polluted. Indiscriminate discharges of industrial effluent, sewage, domestic wastes and remains of human body and other organisms, etc., have not only affected adversely the river water quality and its biota including the fish community, but have also made the river a threat to human health. Due to industrialization in district Hardwar, large numbers of people are immigrating here day-by-day and the old peaceful area is changing into industries and urbanization. Main drain carries away some effluents from the factories and also from the residential colonies. This drains carries domestic sewage, which is poured into river Ganga. Therefore, it can be well considered that it carries a variety of pollutants of equally different Physico-chemical nature. Metallic scraps, metallic dust, oil, paper and wooden products, besides domestic garbage appear to be the main items of the sewage.

It is important that the natural environment of the river should be conductive to the micro and macro fauna so they may thrive, grow and reproduce to the best. On the contrary, the discharge of sewage, domestic wastes, turbidity, siltation and bathing of large number of pilgrims in the water bodies changes the physico-chemical characteristic to such an extent that they sometime lead to high fish mortality. This may be mainly due to the decrease in dissolved oxygen content, increase in biochemical oxygen demand and alkalinity, mechanical injury to the gills, non-availability of the fish food, blocking of migration path, and destruction of spawning ground, etc. The effluents may also adversely affect the development of the fish.

C. Sampling Station of this Study

Following four sampling station were selected for this study, which are located at a distance of 8 km (Fig. 1.2 and Plates 1.1-1.4).

1. *Sampling Station—A (Har-ki-Pauri)*

The stream of the Ganga is separated from Dudhia Dam and flows through Kharkhari to Har-ki-Pauri. At Kharkhari, there is a cremation centre, which is situated on way from Kharkhari to Har-ki-Pauri. Dead bodies are cremated on the right corner of the Baghrao, a seasonal nullah at Kharkhari near Ganga River. The remaining ashes, bones and clothes are thrown into the river Ganga water. Moreover, the sewage from the houses and Ashrams is dumped in the Ganga which pollutes the water. The sewer of Bhimgoda drainage also pollutes the water. Seasonal nullah of Bhimgoda, nullah Baghrao of Kharkhari and nullah near Kangra mandir opens into the river. The current is fast at Har-ki-Pauri but is slower at the opposite Bhimgoda.

Throughout the year, fecal matter, decaying leaves and flowers, wooden parts, waste clothes, food materials and ashes, charcoals can be observed in the Ganga at this point. Hindus come from different parts of the country and release the ashes and bones of dead persons into the Ganga. Each day about 150 kg ashes and bones are thrown at Har-ki-Pauri, which certainly affects the physico-chemical and biological parameters of the Ganga water. About 10,000—15,000 people take bath per day in the Ganga during summer season and about one to two lakhs people take bath per day at Har-ki-Pauri on important bathing festivals. The bottom of the river at this point is stony and sandy.

2. *Sampling Station—B (Mayapuri Ghat)*

The stream of Ganga River flows downward from Har-ki-Pauri to Mayapuri Ghat. This sampling station is at 2 km south

Plate 1.1. Sampling Station A (Har-ki-Pauri)

Plate 1.2. Sampling Station B (Mayapuri Ghat)

Plate 1.3. Sampling Station C (Muneshwar Ghat)

Plate 1.4. Sampling Station D (Pul-Jatwara)

of Har-ki-Pauri. Near to this point is famous Shri Hanuman ji Mandir that is on the west bank of Ganga River. The water current is mild as compare to Har-ki-Pauri. In southern side of this sampling station a small dam control is build.

On all the 365 days of a year, fecal matter, leaves and flowers, wooden parts, waste clothes, food material ashes and charcoals can be observed in the Ganga at this point. This point is used for bathing and washing purpose also. The bottom of the river at this point is stony and sandy.

3. *Sampling Station—C (Muneshwar Ghat)*

This sampling station is between Kankhal and Jwalapur Connection Bridge. The water of this point is polluted by a newly constructed colony on the Northwest banks of river. Many people of this area also pollute the bank of the Ganga by their fecal matter.

The wastewater of the town goes to the side waters of the Ganga. Sewer line of Avas-vikas colony and New Hardwar also pollutes water, which of course will have an impact on physico-chemical and biological parameters. The survey clearly indicates that various sources including decaying clothes, papers etc. are responsible for degrading the general condition of river. At this sampling station the bottom is stony and covered by sand.

4. *Sampling Station—D (Pul Jatwara)*

The down stream of river Ganga flows south towards sampling station Pul Jatwara. This point is located near the small city of Jwalapur. Here flow of stream is little less than Har-ki-Pauri. Width of river is increases as compared to other points.

Entire year fecal matter, decaying leaves and flowers, wooden parts, waste clothes, food material and ashes, charcoals can be observed in the Ganga at this point. The sewer of B.H.E.L. nullah also mixes in the water, which may affect the aquatic ecology of the river. The bottom of river at this point is stony and sandy.

Several workers have made studies on the various aspects of the hydrobiology of different water bodies. Kofoid (1908) studied the plankton of Illionis River. Allen (1920) studied the plankton of the San Joaquin River and its tributaries in California. Abdin (1948a, b) investigated the relationship of physico-chemical factors to algal growth and seasonal distribution of phytoplankton and sessile algae in the river Nile. Anderson *et al.* (1955) noted the phytoplankton—zooplankton relationship in two lakes in Washington. Analysis of the plankton yield in relation to certain physico-chemical parameter of lake Michigan was made by Griffith (1955). Oliff (1959) studied the hydrobiology of the Tugela River. Palmer (1962) investigated the possibility of using algae as bio-indices of pollution. Timms (1970) made a study of the chemical factors and zooplankton of lentic habit in north, eastern New South Wales. Tan (1971) studied the occurrence and ecological adaptation of zooplankton in Tanshi river of Taiwan. Hawkes (1975) made a study of the riverine ecology of Oxford. Holmes and Whitton (1981) made a study of phytoplankton of four rivers, the Tyne, Wear, Tees and Swale.

In India, Pruthi (1933) studied the seasonal changes in the physico-chemical conditions of tank water in the Indian Museum compound. Ghousuddin (1934) made a study of the algal flora of the river Moosi. Ganapati (1940) studied the ecology of the temple tank containing a permanent bloom of *Microcystis aeruginosa.*

After 1950 with growing awakening towards pollutional conditions of our freshwaters bodies, more surveys on different sections of the main rivers and streams have been undertaken in order to access their physico-chemical and biological conditions. Chacko and Krishnamurthy (1954) studied the plankton in ponds of Madras city. Chacko and Srinivasan (1955) made a study on the hydrobiology of the major river of Madras. Rao (1955) examined the distribution of algae in a group of six small ponds. Alikunhi *et al.* (1955) studied the mortality of crop fry in the nursery ponds and role of plankton in their survival growth. Das and Srivastava (1956a, 1956b, 1959) studied the fresh water plankton of tasks and ponds of Lucknow. Chakarabarty *et al.* (1959) made a survey of the plankton and physico-chemical conditions of the Jamuna River at Allahabad. Singh (1960) studied the chlorophyceae of Kumaon.

Limnological studies on the Garhwal Himalayan river were started in 1979 by the pioneering work of Badola (1979). Since then some contributions have been made by Badola and Singh (1981), Dobriyal and Singh (1981), Singh *et al.* (1982), Nautiyal (1990), Dobriyal *et al.* (1983), Pokhriyal *et al.* (1983), Dobriyal (1985) and Dobriyal and Singh (1987).

Various studies have been done to find out physico-chemical, biological and microbial parameter and the inter-relationship of various water bodies in various parts of our country. Some of the important ones are—Gainey and Lord (1957), Kaushik and Prashad (1964), Baloni and Sarkar (1965), Needham and Needham (1972), Badola and Singh (1982), Martin *et al.* (1982), Rai (1984), Bhatt *et al.* (1984), Bhowmick and Singh (1985), Sharma (1985), Belsare (1987), Das *et al.* (1989), Khanna (1993), and Khanna *et al.* (1992, 1997) have studied about physico-chemical and biological parameter of river Ganga at Hardwar. However, the work in detail about the microbial parameters, especially in Bacteria of the river Ganga at Hardwar is scanty. Besides this, regular monitoring of the river Ganga at Hardwar is very important due to its importance as a holy river and increasing instances of pollution.

2

Review of Literature

Whenever pass is reviewed, it shows that beginning of civilization always started or born up only near to the water-sources. A sentence from Yajurveda tells that "anything we try to identify requires water and depends parasitically on water, to this unique indispensable part of our life we pray." Hippocrates (460 BC to 354 BC), the father of medicine, stated that "Water contributes much of health" and asserted that the rain water should be boiled and filtered before it use, otherwise it would have a bad smell and cause harshness (Borchart and Walton, 1971).

In 1971, the first edition of the *Encyclopedia Britannica* defined a filter as a "Stainer" commonly made to bilbous of filtering paper in the form of funnel in order to separate the gross particles from water and render it limpid. Herodotus in his storybook tells about the King of Persia 'Cyrus the great', whenever went for any war, always took boiled water in his silver flagons with him. When observation and results were proved that some water causes disease and infections, then in 1852, a law was enforced in London that, "all potash water henceforth should be filtered". Approximately less than 10 slow sand water filtration plants were commissioned near by 1900.

Later part of 19th century was the period, when people come to know that the relative concentration of ammonia nitrogen, nitrate nitrogen, nitrite and alkanoid nitrogen are pollution indicators of water, by hydro-chemical analysis. In 1921 very first standard criterion for drinking water was reported, in the literature, when United States published USPH standard for drinking water; it specified only bacteriological parameters. The limiting concentrations of three metals (i.e. Lead, Zinc and Copper) were included in 1925 publication. Later in 1946 revision, seven other chemical constituents and in 1962 list of several test were added.

Fresh water lakes and streams together estimate to cover about 0.20 per cent of the earth surface and have a volume of 2.04 per cent x 10^5 km^3 (Lickens, 1961). Although, the surface waters of land in lakes and rivers have received scientist's concentration for functions like navigation, recreation, water supply and fisheries, their ecological aspects have often been ignored. A close analysis of literature in Indian context tells that there is good quantity of valuable data is available of lentic systems in comparison to that of streams and rivers owing to the difficulties in the methodologies to be adopted in river system. Since 1960, investigation on the ecology of river has gained value and the researches were carried out hitherto mainly concentrated on aspects such as physico-chemical nature of water, phytoplankton, zooplankton, macrophyte study, productivity, heavy metals level and microbial ecology.

However, it is difficult to give the name of scientist, Limnologist and Hydrobiologist who worked on the various aspects of water quality and aquatic ecology. This review includes some of the important contribution made from various parts of the world. Many workers from India and abroad discussed the different aspects of water quality time to time. Hence, some brief references are being quoted here.

Kofoid (1908) studied the plankton of Illinois river. Allen (1920) studied plankton of the San Joaquin river and its tributaries in California. Richardson (1921) reported various changed in the composition of bottom and shore fauna of Illinois river and its connecting lakes. Clarke (1924) analysed the composition of river and lake water of United State. Galtsoff (1924) and Reinhard (1931) studied the limnology of upper Mississippi. Greenfield (1925) observed the chemical and biological parameters of Illinois river. Eddy (1934) studied the fresh water plankton communities of Illinois river. Chandler

(1940) made limnological studies of western lake Eric. Velz (1947) explain the process of recovery and self-purification of river water. Abdin (1948a, b) investigated the relationship of physico-chemical factors to algal growth and seasonal distribution of plankton and sessile algae in the river Nile. Berner (1951) made a study about limnology of the lower Missouri river. Anderson *et al.* (1955) noted the phytoplankton-zooplankton relationship in two lakes in Washington. Griffith (1955) analysed the plankton yield in relation to certain physico-chemical parameters of lakes Michigan. Oliff (1959) analysed the hydrobiology of the Tugela river. Holden and Green (1960) studied the hydrology and plankton of the river Sokoto. Ward and Seibert (1963) made studies on the limnology of Dow lake. Dorris (1963) studied the limnology of the middle Mississippi. Train (1967) evaluated quality criteria for water. Hynes (1970) studies the ecology of running water; Cosey and Newton (1973) discussed the chemical composition and flow of the river Frome and its tributaries. Hawkes (1975) made a study of the river ecology of Oxford. Robinson and Kaller (1976) made a comparative study on the water characteristics of four northern-west Virginia rivers. Horned (1982) analysed the water quality of Neuse river in terms of variability pollution load and long-term trend. Daniel *et al.* (1982) analysed the water quality of the French Broad river, based on the data collected at Marshall between 1958-77. Probost (1986) examined dissolved and suspended matter transported by Girau river in France.

Atkin and Harris (1924) observed seasonal changes and Heleoplankton of fresh water ponds. Alexander *et al.* (1935) a survey of the river Tees. Garner *et al.* (1936) observed the chemical and biological aspects of river Holme. Ellis *et al.* (1946) made a study of the determination of water quality. Dorris (1963) studied the limnology of the midle Mississippi. Freiser and Fernado (1966) observed the ionic equilibria in analytical chemistry. Crayton and Sommerfield (1979) analysed the composition and abundance of phytoplankton in the tributaries of the lower Colarado river. Whitton (1983) studied the ecology of a number of European rivers.

Many workers of India made various discoveries and studies on the rivers, some of them are as follows:

Ghosuddin (1934) made a study of the algal flora of the river Mossi. Ganpati (1940) studied the ecology of the temple tank containing a permanent bloom of *Microcystic aeruginousa.* Ganpati and Alikunhi (1950) observed that water of the river Ganga was heavily polluted, which was dangerous to the health of the people. Chakarbarty *et al.* (1959) made a survey of the plankton and physico-chemical conditions of the Jamuna river at Allahabad. Singh (1960) studied the *Chlorophyceae* of Kumaon. Verma (1964) carried out hydrobiological survey of Dalsagar tank. Rao and Govind (1964) carried out the hydrobiology of Tungbhadra reservoir. Baloni and Sarkar (1965) made some observation on the pollution of Yamuna river at Okhla water works intake Delhi. Saxena *et al.* (1966) observed pollution of the river Ganga near Kanpur. Kant and Kachroo (1973) studied the limnology of Kashmir lakes. Aggarwal *et al.* (1976) analysed the hydrobiology of river Ganga at Varanasi. Kant and Anand (1978) investigated the inter-relationship of phytoplankton and physical parameters in Mansar lake of Jammu. Rai (1978) made ecological studies of algal community of river Ganga at Varanasi.

Badola and Singh (1981) studied hydrobiology of the Alaknanda. Gopal *et al.* (1981) studied the limnology of a fresh water reservoir of Jaipur. Raina *et al.* (1984) made pollution studied of river Jhelum. Chattopadhya *et al.* (1984) made a study on the pollution status of river Ganga at Kanpur. Bhatt *et al.* (1985) explain the ecology and phytoplankton population in relation to physico-chemical factors of river Kosi of western Himalaya. Sharma (1986) published a report on effect of physico-chemical factors and benthic fauna of Bhagirathi river Garhwal Himalaya.

Nautiyal (1990) explain the ecology of the Ganga river system in the upland of Garhwal Himalaya environment. Pathak (1991) studied some aspects of ecology of the rivers Sarju and Gomti of the Kumoau Himalaya. Singh (1991) published a book "Advances in limnology". Mazhar and Kapoor (1992) made a limnological study on Doraniu river at Bareilly. Joshi *et al.* (1993) evaluated the physico-chemical characteristic of river Bhagirathi in the uplands of Himalaya Khanna *et al.* (1993) studied about physico-chemical and Biological parameter of river Ganga at Chandighat, Hardwar. Pandey *et al.* (1993) studied on the physico-chemical quality of water or river Kosi at Purnia. Kumar (1995) studied on pollution in river Mayurakshi in south Bihar. Shivsubramani and Mahadevan (1995) evaluated the water quality of river Suruliyar in Tamil Nadu. Joshi *et al.* (1995) analysed primary productivity in western Ganga canal at Hardwar. Shrivastava *et al.* (1996) evaluate the phytoplankton productivity and physico-chemical properties of Repti river. Joshi (1996) studied the hydro biological profile of river Sutlej in its middle stretch in western Himalaya.

Venkateswarlu and Jayanti (1968) studied the hydrobiology of the river Sabarmati. Mohanty (1981) studied the water quality of some water bodies of Bhubaneswar. Verma *et al.* (1984) observed the pollution and Saprobic status of eastern Kalinadi. Bhatt *et al.* (1985) studied the hydrobiology and phytoplankton population in river Kosi. Ramesh *et al.* (1992) evaluated the water quality in three rivers of Andhra Pradesh. Saxena and Chauhan (1993) observed the physico-chemical aspects of pollution in river Yamuna at Agra. Jameel (1998) made a physico-chemical study in Uyyakondan channel water of river Cauvey. Sharma and Pandey (1998) investigated the pollution status on Ramganga river at Morababad.

Chopra and Patric (1994) studied the effect of domestic sewage on self-purification of Ganga water at Rishikesh. Chopra and Rehman (1995) made a study on self-purification of physico-chemical properties of Ganga canal water at Jwalapur, Hardwar Joshi *et al.* (1996) observed the planktonic population in relation to certain physico-chemical factors of Ganga canal at Jwalapur, Hardwar. Khanna *et al.* (1997) observed the population of Green algae in relation to physico-chemical factors of the river Ganga at Hardwar.

Following are some other workers worked on Indian rivers extensively for physico-chemical aspects. Chacko and Ganapati (1949) on the river Adyar; Deshmukh *et al.* (1964) on the river Kanhan; George *et al.* (1966) on the river Periyar; Ray *et al.* (1966), and Bilgrami and Duttamunshi (1985) on the river Ganga; Badola and Singh (1981), Aggarwal (1986), and Nautiyal *et al.* (1986) on the river Alakananda; Somashekar (1984) on the river Cauvery and Reddy and Venkateswarlu (1987) on the river Tugabhadra provided valuable information on physico-chemical data.

Moreover, physico-chemical factors, many of scientist worked on pollution aspects of rivers and investigated in this direction are Motwani *et al.* (1956) on the river Sone; Aggarwal *et al.* (1986) on the river Chambal; and Sivakumar *et al.* (1987) on the river Amaravati. Much of the data compiled in these studies were on different aspects of pollution and the impact of pollutants on the biota of river systems. Ghosh and Basu (1968) while working in Hooghly estuary reported that the waste discharges from a chemical factory contain both inorganic and organic pollutants, which highly effect on the downstream of the discharge point as compared to upstream points. Due to high dilution and heavy mixing the estuary recovers from the hazards of pollution, as it travels at a distance of about 3.5 km downstream.

Many Indian rivers were studies in the last four decades on the algal taxonomy. Notable among them are Iyengar and Venkataraman (1951) for the river Cooum, Lakshminarayanan (1965). Some of the above works revealed the role of various chemical parameters in the seasonal dynamics of phytoplankton. Zafar (1967) showed that organic matter played an important role in the influence on phytoplankton population and a more or less inverse relationship with oxygen concentration and pH. Singh (1967) found no such correlation between pH and *Myxophyceae* populations with an increase in pH. Jackson (1961) reported that the optimum range of alkalinity (50-110 ppm) for the blooming the blue-green algae.

Zafar (1967) found that among the *Chlorophyceae,* the desmids thrive well in higher basis ratios of sodium and potassium to calcium and magnesium (1.2-3.4 mgl^{-1}) but are highly susceptible to higher concentrations of the calcium, disappearing totally when the concentration of calcium exceed 40 mgl^{-1}. Ray (1955) established a direct correlation between salinity and diatom population and also temperature and turbidity mainly influence phytoplankton of the river Hooghly. Temperature influences the time of occurrence of different phytoplankton while turbidity affects the plankton abundance. Thus diatoms were reported to have some minimum, optimum and maximum temperatures, viz., below 20°C and 23°C, 20°-23°C and 31°-32°C respectively. Similarly, *Oscillatoria agardhii* and

O. pseudogeminata occur within different ranges of temperature, viz 24°-32°C and 23°-29°C respectively. Similarly the genera belonging to Chlorococcales and Conjugales on the one hand and *Geminella interrupta* on the other occur within different temperature ranges of 25°-31°C and 20°-21°C respectively.

Turbidity appears to have an inverse influence on the total phytoplankton crop. A similar correlation between turbidity and low basic productivity of the Missouri River was observed by Berner (1951). Butcher (1946-49) however was of the opinion that current probably helps in the growth of the microalgal flora of a river. Thus, a number of factors were reported to affect the seasonal growth of phytoplankton.

Trace metal levels in river systems were studied at different component levels. River water of Cauvery, Damoder, Ganga, Kapila, Moosi, Purna and Ulhas were studied for various heavy metals. All these studies indicate that zinc was most commonly occurring element with a range (0.05-25.5 ppm) in all the above river systems. The least occurring metals include arsenic, cadmium and mercury.

Sediment of different rivers was examined with respect to heavy metals. The sediments of Damodar, Hooghly and Ulhas exhibit significant amount of zinc. Analysis of the various components revealed that sediments accumulate significant amount of these metals.

Many of the species were recognized as bioindicators for specific pollutants. Patrick (1978) used algae as indicators of pollution. Cairns and Lanza (1972) found both algae and ciliates as useful indicators of water quality. Jeejabai and Rajendran (1982) used changing pattern of plankton composition as indicator of water pollution. Govindan and Sundaresan (1979) examined seasonal succession of algae in relation to pollution levels in Adyar river. Prasad and Yashpal (1982) emphasized diatoms as indicator of pollution. There are also studies on the toxic effects of heavy metals on fresh water algae. Bharati *et al.* (1979) noticed physiological imbalance in fresh water algae due to hexavalent chromium. Fesher *et al.* (1984) studied the accumulation and toxicity of Cd, Zn, Ag and Hg in four species of algae namely *Thalassiosira pseudonana, Dunaliella tertiolecta,*

Emiliania hyxleyi and *Osillatoria woronichinii* and found that dead cells of these species had accumulated metals better comparable to living cells. They inferred that initial association of metal with the cells in governed by mere absorption. Gupta (1989) reported the interactive effects of inorganic nitrogen nutrients and copper on the growth of bloom forming blue-green alga *Microcystis aeruginosa*. Ho (1984), Welch (1984) and Whitton (1970, 1985) carried out investigations on heavy metals in aquatic ecosystems concerned either with toxicity or accumulation, especially on algae.

3

Materials and Methods

For microbial ecology study of Ganga water at Hardwar the water samples were collected seasonally from different sampling stations (A, B, C, and D) during April 1998 to March 2000 in morning hours (from 7.00 AM to 10.00 AM). The samples were taken in borosil glass bottles of 300 ml, plastic cans of 1 lt. from each location.

The samples for different parameters were analysed with the help of the procedure described by Welch (1948), APHA (1980), Mathur (1982), Ross (1983), Trivedi and Goel (1984) and Khanna (1993).

Sampling Method

A. Cleaning and Sterilization of Glasswares

For cleaning the glass wares washing soda or chromic acid was used.

1. **Sample Battle:** 300 ml capacity BOD bottles made of borosil were used. They were washed with washing soda or chromic acid and rinsed with tap water followed by distilled water. The neck and stopper were wrapped by butter paper with the help of rubber band. Sterilization of the sample bottle was done in autoclave at 15 lbs/inch^2 pressure and 121°C for 20 minutes.

2. **Pipettes:** Different volume size pipettes were washed and fitted with cotton plug at the upper end. These were wrapped in butter paper and sterilization was

done in autoclave at 15 lbs/inch^2 pressure and 121°C for 20 minutes.

3. **Petridishes:** Borosil 90 mm diameter petriplates were washed and then sterilization was done in an oven at 180°C for 2 hours.

4. **Test Tubes:** Borosil test tubes were washed and then plugged with non-absorbent cotton wool. These were arranged in test tube racks and sterilization was done in autoclave at 15 lbs/inch^2 pressure and 121°C for 20 minutes.

B. Sampling

For physico-chemical and microbiological parameters, Ganga river water sample was collected from four sites in plastic cans. Sampling for dissolved oxygen (DO) and bacteriological analysis was done separately.

1. **Sampling for Dissolved Oxygen (DO):** The samples were collected in clean and sterilized 300 ml capacity BOD bottles. The bottles were filled completely with Ganga water and stoppered was placed inside the water only. Immediately DO was fixed by adding 2 ml of each alkaline KI and manganous sulphate ($MnSO_4$) at the sampling site.

2. **Sampling for bacteriological analysis:** Cleaned and sterilized sampling tubes was used. It was dipped into the water with its neck downward upto 6 inches below the water surface and then its mouth was opened under water against the flow. The tube was not filled completely and about 10% of the volume of the tube was kept empty. It was closed immediately taking out of the water. To avoid contamination of sample, the neck and the stopper were wrapped by butter paper with the help of rubber band.

C. Water Anlysis Methods

The physico-chemical and microbiological analysis of water samples were made by the following methods out lined in Welch (1948), Mark (1977), APHA (1980), Kudesia (1980), Mathur

(1982), Trivedi and Goel (1984) and Khanna (1993). Samples are analysed for the following parameters.

I. Physico-chemical Parameters

1. Temperature
2. Turbidity
3. Conductivity
4. Velocity
5. Total solids (TS)
6. Total dissolved solids (TDS)
7. Total suspended solids (TSS)
8. pH
9. Free Carbon dioxide (CO_2)
10. Dissolve oxygen (DO)
11. Biochemical oxygen demand (BOD)
12. Chemical oxygen demand (COD)
13. Total Alkalinity
14. Chloride (Cl^-)
15. Total Phosphate
16. Total organic carbon
17. Hardness
18. Ions:

(a) Sodium (b) Potassium (c) Sulphate (d) Nitrate

II. Microbial Parameters

19. Heavy Metals
20. Standard Plate Count (SPC)
21. Most probable number (MPN) for total coliforms
22. Plankton

Temperature, pH, CO_2, DO, BOD, SPC and MPN were analysed immediately after sampling.

Methodology

1. ***Temperature***

This parameter is basically important for its effects on the chemistry and biological reactions of the organisms in the water.

The thermometer is the best suitable device for measuring the temperature. The bulb of thermometer dipped into the water and the reading was noted in °C (Celsius).

2. ***Turbidity***

Turbity in water is caused by the substances not present in the form of true solution. True solutions have a particle size of less then 10^{-9} m. Any substance having more than this size will produce a turbidity. Turbidity of water is actually the expression of optical property (Tyndall effect) in which the light is scattered by the particles present in the water.

Turbity in natural waters is casused by clay, silt, organic matter, phytoplankton and other microscopic organisms. Turbity makes the water unfit for domestic purpose, food and beverage industries, and many other industrial uses.

The turbidity of the water was estimated by Jackson's candle turbidimeter. The water samples were collected from the river in plastic cans; collected sample were shaked well and transferred into the glass-tube of Jackson's candle turbidimeter until the flame of the candle disappears and noted the light path in centimeter (measured from inside the bottom of the glass tube), and the reading recorded in JTU.

3. ***Conductivity***

Electrical conductance is the ability of a substance to conduct the electric current. In water, it is the property caused by the presence of various ionic species.

Conductivity of the sample was determined by the ELICO conductivity metre.

4. ***Velocity***

The speed of water current was measured by the surface float method. The rate of flow of the uppermost stratum of the water in the river was measured by throwing a light piece of wood in different segments of the river. Then the time taken by the wood piece to cover a known distance was calculated by the given formula:

V = d/t

Where, V = velocity (m/sec)

d = distance covered by the float (metres)

t = time taken by the float (sec.).

5. ***Total Solids (TS)***

Total solids or total residue is the term applied to material left in a beaker after evaporating a well mixed sample and subsequently evaporating and dried it in a oven.

Procedure

A 100 ml washed and dried beaker was taken and weighed. 100 ml of unfiltered sample in the beaker, was evaporated on water bath then dried it in an oven at 105±1°C. The final weight of the beaker was taken after drying of sample.

Calculation:

$$\textbf{Total solids, mg/L} = \frac{\textbf{A-B} \times \textbf{10}^6}{\textbf{V}}$$

Where, A = Final weight of beaker in g.

B = Initial weight of beaker in g.

V = Volume of the sample taken in ml.

6. ***Total Dissolved Solids (TDS)***

Total dissolved solids or filterable residue are those solids, which left after evaporation of the filtered sample.

Procedure

A 100 ml washed and dried beaker was taken and weighed immediately before use. The filter paper Whatman no. 42 was placed in funnel beaker and the 100 ml of well-mixed sample was filtered through filter paper. The filtrate was collected in the 100 ml-weighed beaker. After complete filtration, the sample was evaporated and subsequently dried in an oven at 105±1°C for 4 to 6 hrs. Finally the beaker was cooled and weighed.

Calculation:

$$\textbf{Total dissolved solids, mg/L} = \frac{\textbf{A-B x } \mathbf{10^6}}{\textbf{V}}$$

Where, A = Final weight of beaker in g.

B = Initial weight of beaker in g.

V = Volume of the sample taken in ml.

7. ***Total Suspended Solids (TSS)***

Total suspended solids or non-filterable residue are the retained material on Whattman no. 42 filter paper after filteration of well-mixed sample.

Procedure

TSS was determined by taking difference between the total solids and total dissolved solids.

Calculation:

TSS, mg/L= TS-TDS

8. ***pH (Potentia hydrogenii)***

pH is the hydrogen ion activity or concentration in the given water sample. pH equals to negativity $\log_{10}$ of hydrogen ion concentration.

$$pH= -\log_{10} (H^+)$$

pH can be determined by the pH strips, but the accurate method is the electrometric method in which potential in between the two electrodes (indicator and calomel) is measured directly. pH of the Ganga river water sample was measured with the help of electrodes of pH metre (ELICO). The apparatus has a reference and an indicator glass electrode.

Procedure

The apparatus was calibrated with the help of buffer solutions having known pH 4.01 and 9.18 called as pathalate and borate buffers respectively. Now both of the electrodes were rinsed with distilled water and then dipped in the test solution taken in a 100 ml-glass beaker. As the electrodes were dipped in the sample, the pH metre gives the value of the test solution.

Reagents

(a) Potassium hydrogen pathalate buffer:

10.2 g of potassium hydrogen pathalate was dissolved in distilled water to prepare 1000 ml of buffer.

(b) Borax buffer:

3.81 g of $Na_2B_4O_7.10H_2O$ was dissolved in distilled water to prepare 1000 ml of buffer.

9. Free Carbon Dioxide (CO_2)

Principle

Free CO_2 can be determined by titrating the sample using a strong alkali (such as carbonate free NaOH) to pH 8.3. At this pH all the free CO_2 is converted into bicarbonates.

Reagents

(a) Sodium hydroxide, 0.05 N:

1.0N NaOH was prepared by dissolving 40 g of NaOH in CO_2 free distilled water (boiled) to make 1.0 litre of solution. 50 ml of 1.0N NaOH was diluted to 1 litre. Standardization of the solution was done with H_2SO_4.

(b) Phenolphthalien indicator:

0.5 g of Phenolphthalein was dissolved in 50 ml of 95% ethanol and 50 ml of distilled water was added. Drop by drop 0.05 N CO_2 free NaOH solution was added drop wise, until the solution was turned faintly pink.

Procedure

100 ml of sample was taken in a conical flask and a few drops of phenolphthalein indicator was added. If the colour turns pink, free CO_2 declared to absent. If the sample remained colourless, titrated it against 0.05N NaOH. At the end point a pink colour was appeared.

Calculation:

$$\textbf{Carbon dioxide } (CO_2), \textbf{ mg/L} = \frac{(\text{ml} \times \text{N}) \text{ of NaOH} \times 1000 \times 44}{\text{ml of sample}}$$

10. *Dissolved Oxygen (DO)*

DO is the measure of oxygen concentration present in a given water sample. The concentration of oxygen reflects whether the process undergoing is aerobic or anaerobic.

There are two ways for the detection of the DO.

A. *Winkler's Idometric method*

B. *Azide modification of Winkler's method*

Although there is minute change between the two methods, but we followed the second, most suitable for Ganga Water Samples.

AZIDE MODIFICATION OF WINKLER'S IODOMETRIC METHOD

Principle

The manganous sulphate reacts with the alkali (KOH or NaOH) to form a white precipitate of manganous hydroxide, which in the presence of oxygen gets oxidized to brown colour compound. In the strong acid medium, manganic ions are reduced by iodide ions, which get converted, to iodine equivalent to the original concentration of the oxygen in the sample. The iodine can be titrated against this sulphate using starch as indicator.

Certain oxidizing and reducing materials are effectively interfere with the determination of oxygen by converting iodide on ions to iodine or vice versa. Azide modification removes such substance specially nitrite, which is distorted by sodium azide (NaN_3).

Reagents

(a) *Sodium thiosulphate, 0.25N:*

24.82 g of $Na_2S_2O_3 \cdot 5H_2O$ was dissolved in boiled distilled water and makeup the volume to 1 lt. 0.4 g of borax or a pallet of NaOH was added as stabilizer. This was 0.1 N stock solution. It was diluted 4 times with boiled distilled water to repair 0.25 N solution. It was kept in a brown glass stoppered bottle.

(b) Alkali iodide azide solution:

(i) 500 g of NaOH or 700 g of KOH and 150 g of KI was dissolved to make 1 lt. of solution. (ii) 10 g of NaN_3 was dissolved in distilled water. Two solution (i) and (ii) were mixed.

(c) Manganous sulphate solution:

100 g of $MnSO_4 \cdot 4H_2O$ was dissolved in 200 ml of boiled distilled water and filtered.

(d) Starch solution:

1 g of starch was dissolved in 100 ml of warm distilled water and a few drops of formaldehyde solution was added.

(e) Sulphuric acid:

H_2SO_4, conc. (sp. Gr. 1.84)

Procedure

The sample was filled in a glass stoppered bottle (BOD bottle) of 300 ml capacity, avoiding any kind of bubbling and trapping of air bubble after placing the stopper. 2 ml of each $MnSO_4$ and alkaline KI were poured into the bottom of sample by the pipette. A precipitate was appeared. The stopper was placed and shake the contents well by inverting the bottle repeatedly. The bottle was kept for some times. This is the stage, where the DO was fixed. Now 2 ml of conc. H_2SO_4 was added and shook to dissolve the precipitate; and titrated the whole content against hypo solution using starch as an indicator, till the blue colour changed to colourless. Now the DO was calculated.

Calculation:

$$\textbf{Dissolved oxygen, mg/L} = \frac{\textbf{(ml x N) of titrant x 8 x 1000}}{\textbf{V1-V}}$$

Where, V1 = Volume of bottle after placing the stopper

V = Volume of $MnSO_4$ and KI added.

11. *Biochemical Oxygen Demand (BOD)*

BOD is the measure of biodegradable organic matter present in a water sample and can be defined as the among of oxygen required by the microbes in stabilizing the biologically degradable organic matter under aerobic conditions. Thus BOD value can be used as a measure of waste strength.

Principle

The principle of the method involves, measuring the difference of the oxygen concentration between the sample and after incubating it for 5 days at 20°C.

Apparatus and Reagents:

(a) *BOD bottles*

(b) *BOD incubator*

(c) *Phosphate buffer:*

Each 8.5 g of KH_2PO_4, 21.75 g K_2HPO_4 33.4 g $Na_2HPO_4 \cdot 7H_2O$ and 1.7 g NH_4Cl was dissolved in distilled water to prepare 1 litre of solution. pH was adjusted to 7.2.

(d) Magnesium sulphate:

82.5 g $MgSO_4 \cdot 7H_2O$ was dissolved in water to prepare 1 litre of solution.

(e) Calcium chloride:

27.5 g of anhydrous $CaCl_2$ was dissolved in distilled water to prepare 1 litre of solution.

(f) Ferric chloride:

0.25 g $FeCl_3 \cdot 6H_2O$ was dissolved in distilled water to prepare 1 litre of solution.

(g) Sodium thiosulphate, 0.25N:

24.82 g of $Na_2S_2O_3 \cdot 5H_2O$ was dissolved in boiled distilled water and makeup the volume to 1 lt. 0.4 g of borax or a pallet of NaOH was added as stabilizer. This was 1 N stock solution. It was diluted 4 times with boiled distilled water to repair 0.25 N solution. It was kept in a brown glass stoppered bottle.

(h) *Alkali iodide azide solution:*

(i) 500 g of NaOH or 700 g of KOH and 150 g of KI was dissolved to make 1 lt. of solution. (ii) 10 g of NaN_3 was dissolved in distilled water. Two solution (i) and (ii) were mixed.

(i) *Manganous sulphate solution:*

100 g of $MnSO_4 . 4H_2O$ was dissolved in 200 ml of boiled distilled water and filtered.

(j) *Starch solution:*

1 g of starch was dissolved in 100 ml of warm distilled water and a few drops of formaldehyde solution was added.

(k) *Sulphuric acid:*

H_2SO_4, conc. (sp. Gr. 1.84)

Procedure

Dilution water was prepared by bubbling compressed air in distilled water for about 30 minutes. Now 1 ml each of $MgSO_4$, $FeCl_3$, PO_4 buffer and $CaCl_2$ was added into one litre of water and mixed thoroughly. The sample was neutralized around pH-7.0. As the DO in the sample is likely to be exhausted, so a suitable dilution of the sample was prepared according to the expected BOD range. This dilution sample water filled the two sets of BOD bottles. One set of the bottles was kept in BOD incubator at 20°C for 5 days and the DO constant was determined in another set immediately. After the compilation of 5 days incubation, the DO of first set was determined. BOD was calculated by the given formula.

Calculation:

BOD, mg/L= (DO-D5) x dilution factor

Where, DO = Initial DO in the sample

D5 = DO after 5 days.

12. *Chemical Oxygen Demand (COD)*

COD is widely used to characterize the organic strength of wastewaters and pollution of natural waters. The test measures the amount of oxygen required for chemical oxidation of organic matter in the sample to carbon dioxide and water.

Principle

The sample is refluxed with $K_2Cr_2O_7$ and H_2SO_4 in presence of $AgSO_4$ (catalyst) and $HgSO_4$ (to neutralize the effect of chlorides). The excess of $K_2Cr_2O_7$ is titrated against ferrous ammonium sulphate using ferroin as an indicator. The amount of $K_2Cr_2O_7$ used is proportional to the oxidizable matter present in the sample.

Reagents

(a) *Potassium dichromate solution, 0.025 N:*

12.259 g of dried A.R. grade $K_2Cr_2O_7$ was dissolved in distilled water to make 1 litre of solution. This is 0.25N $K_2Cr_2O_7$ solution. It was diluted 10 times to make 0.025N $K_2Cr_2O_7$ solution.

(b) *Ferrous ammonium sulphate, 0.01N:*

39.2g of $FeSO_4$ $(NH_4)_2SO_4.6H_2O$ was dissolve in water, adding 20 ml conc. H_2SO_4 to make 1 litre of 0.1 N solution. It was diluted 10 times to (100→1000 ml) to make 0.01 N $FeSO_4(NH_4)2SO_4.6H_2O$ solution.

(c) *Ferroin indicator:*

1, 10-phenonthroline and 0.695 g of ferrous sulphate ($FeSO_4.7H_2O$) were dissolved in distilled water to make 100 ml of solution.

(d) *Sulphuric acid:*

H_2SO_4, conc. (sp. gr. 1.84).

(e) *Mercuric sulphate:*

$HgSO_4$, solid.

(f) *Silver sulphate:*

Ag_2SO_4, solid.

Procedure

20 ml of sample was taken in COD flask. 10 ml of 0.025 N $K_2Cr_2O_7$ and a pinch of Ag_2SO_4 and $HgSO_4$ were added. (In case

of the COD is expected below 50 mg/L, 10 ml of 0.025 N $K_2Cr_2O_7$ should be added.) Now 30 ml of conc. H_2SO_4 was added and reflux condenser was connected to the COD flask. It was refluxed for 2 hours on water bath. The flask was removed, cooled and distilled water was added to make the final volume to about 140 ml 2-3 drops of ferroin indicator was added and titrated with 0.1 N $FeSO_4$ (NH_4) $_2SO_4$.

A blank was also run with distilled water, using same quality of chemicals.

Calculation:

$$\text{COD, mg/L} = \frac{\text{(B-A) x N of } FeSO_4 \ (NH_4) \ 2SO_4 \text{ x 1000 x 8}}{\text{Volume of sample (ml)}}$$

Where, A = Volume of titrant with sample (in ml).

B = Volume of titrant with blank (in ml).

13. *Total Alkalinity:*

It is the measure of the capacity of the water to neutralize a strong acid.

Principle

Total alkalinity can be estimated by titrating the sample with a strong acid (HCl or H_2SO_4), first to pH 8.3 using phenolphthalein as an indicator and then further to pH between 4.2 and 5.4 with methyl orange. In the first case the value is called as phenolphthalein alkalinity (PA) and in second case, the value is called as total alkalinity (TA). Values of carbonates, bicarbonates and hydroxyl ions can be computed from these two types of alkalinities.

Reagents

(a) *Hydrochloric acid, 0.1N:*

12 N concentration HCl (sp. Gr. 1.18) was diluted to 12 times (8.34→100 ml) to prepare 1.0 N HCl. It is further diluted ten times to make 0.1 N HCl (100→1000 ml).

(b) *Phenolphthalein indicator:*

0.5 g of Phenolphthalein was dissolved in 50 ml of 95% ethanol and 50 ml of distilled water was added. Drop by drop

0.05 N CO_2 free NaOH solution was added drop wise, until the solution was turned faintly pink.

(c) Methyl orange indicator, 0.05%:

0.05 g of methyl orange was dissolved in 100 ml of distilled water.

Procedure

100 ml of sample was taken in conical flask and 2 drops of phenolphthalein were added (i) solution remained colourless, so phenolphthalein alkalinity was absent (PA = 0) and total alkalinity was determined. (ii) If the colour changes to pink, it is titrated with 0.1 N HCl until the colour disappeared at the end point. This is phenolphthalein alkalinity (PA).

Now, 2 drops of methyl orange were added to the same sample and titration was continued with 0.1 N HCl, until yellow colour changed to pink. This was total alkalinity (TA).

Calculation:

$$\textbf{PA as CaCO}_3\textbf{, mg/L} = \frac{\textbf{(A x normality of HCl) x 1000 x 50}}{\textbf{Volume of sample (ml)}}$$

$$\textbf{TA as CaCO}_3\textbf{, mg/L} = \frac{\textbf{(B x normality of HCl) x 1000 x 50}}{\textbf{Volume of the sample (ml)}}$$

Where, A = ml of HCI used with only phenolphthalein

B = ml of Total HCl used with only phenolphthalein and methyl orange.

14. *Chloride (Cl^-):*

The most important source of chloride in the water is industries and discharge of domestic sewage. The chloride concentration serves as an indicator of pollution by sewage.

Principle

Silver nitrate reacts with chloride to form very slightly soluble white precipitate of silver chloride (AgCl). At the end point where all the chlorides gets precipitated, free silver ions react with chromate to form silver chromate of reddish brown colour.

Reagents

(a) Silver nitrate, 0.02N:

3.4 g of dried $AgNO_3$ (A.R.) were dissolved in distilled water to make 1 litre of solution and kept in a dark bottle.

(b) Potassium chromate 5%:

5 g of K_2CrO_4 was dissolved in distilled water.

Procedure

50 ml of sample was taken in an Erlenmeyer flask and added 2 ml of K_2CrO_4 solution. The contents were titrated against 0.2 N $AgNO_3$ until a red tinge appeared.

Calculation:

$$\text{Chloride, mg/L} = \frac{(\text{ml} \times \text{N}) \text{ of } AgNO_3 \times 1000 \times 35.5}{\text{Volume of sample (ml)}}$$

15. *Total Phosphate*

50 ml of sample was taken in a flask followed by the addition of acid ammonium molybedate solution (2 ml) and 4-5 drops of $SnCl_2$ solution. A blue colour appeared. The same procedure was run with a blank (distilled water) sample for comparison. Both (sample and blank) were then kept on white paper, followed by the addition of standard phosphate solution to the blank drop by drop with the help of graduated pipette (1 ml) until the colour of blank reached at equivalent to the sample, at the end point the ml of standard phosphate used was placed in the equation to get the amount of phosphate.

16. *Total Organic Carbon*

Measuring Principle (Shimadzu)

Measurement of TC (Total Carbon)

TC combustion tube is filled with oxidation catalyst and heated to 680°C. Carrier gas (high purity air) is supplied into this tube after it is controlled at a flow rate of 150 ml/min. By a pressure controller and a mass flow controller and moistened by a humidifier. When sample has been introduced by a sample injector into the TC combustion tube (100 µl at the maximum for TC catalyst and 2,000 µl at the maximum for high sensitivity TC catalyst), TC component in the sample combusted or decomposed to become CO_2. The carrier gas which contains combustion tube flows through an IC reaction vessel and cooled and dried by a dehumidiffer. It is then sent through a halogen scrubber into a sample cell set in a non-dispersive infrared gas analyser (NDIR) where CO_2 is detected. The NDIR outputs a detection signal (analog signal) which generates a peak whose area is calculated by a data processor.

The peak area is proportional to the TC concentration of the sample. Therefore, if calibration curve equation expressing the relation using TC standard solution, the TC concentration of the sample can be determined from the calculated peak area.

For reference, TC is composed of TOC (total organic carbon) and IC (inorganic carbon).

Measurement of IC (Inorganic carbon)

Sample (2,000 µl at the maximum) is introduced with a sample injector into an IC reactor vessel where carrier gas is flowing in form of tiny bubbles in the solution is decomposed to become CO_2 which is detected by the NDIR. The IC concentration can be determined in the same procedure as the TC concentration.

Carbon in form of carbonates and hydrogen carbonates can be measured as IC.

Measurement of TOC (Total organic carbon)

TOC concentration can be obtained by subtracting the IC concentration from the TC concentrated as above.

17. Hardness

Total hardness was determined by using EDTA titrametric method: A 100 ml sample of river water was titrated by 0.01N EDTA solution after the addition of indicator (very small amount of EBT + 1 ml of amonium buffer). End point was recorded at the time of first colour change (pinkish to blue) and total hardness was then calculated by using the formula.

Calculation:

$$\text{Hardness (mg/l)} = \frac{\text{ml of EDTA used x 1000}}{\text{ml of sample}}$$

18. IONS:

(a) Sodium (b) Potasium (c) Sulphate and (d) Nitrara-By AAS

19. Heavy Metals-By Atomic Absorption Spectrophotometry (AAS) (Shimadzu)

Principle of AAS

AAS consisting of light source emitting the line spectrum of element (hallow catode lamp or electrode less discharge lamp), a device for evaporizing the sample (using a flame, a means of isolating an absorption line, monochromator or filter and adjustable slit) and a photoelectric detector with its associated electronic amplifying and measuring equipment.

Method of using AAS

Requirements for determining by atomic absorption spectrophotometry (AAS) vary with and/or concentration to be determined. In atomic absorption spectrophotometery, a sample is aspored into a flame and atomized and a lightbeam is directed through the flame, into a monochrometer and on to a detector that measure the amount of light absorbed by the atomized element in the flame. Each metal has its own characteristic absorption wavelength. A source lamp composed of that element is used. The amount of energy of the characteristic wavelength absorbed in the flame is propotional to the concentration of element in the sample. For the detection of sodium and potasium the sample was directly aspirated into air acetylene flame and the concentration recorded.

The instrument was calibrated, for each metal to be detected, with known standard. After calibration and programming, the prepared samples were aspirated to the flame. Zero was set with double distilled water before each determination. Results are displayed in μgm/l from which the actual concentration in water samples was calculated. It was the total content of water sample, which was expressed in mg/l (ppm).

20. *Standard Plate Count (SPC)*

On a solid medium, counting of organisms depends upon the fact that living cells will produce sufficient progeny to form a colony visible to necked eyes. Since bacteria occurring in water are single cell, pairs, groups or even dense clumps, not every individual living cell will develop into a separate colony on incubation. Therefore a number of colonies appearing on plate does not necessarily represent the total number of organisms present in the test volume. The results are expressed as the cell forming units (CFU) per ml.

Apparatus and Materials

(a) Incubator

(b) Petridishes

(c) Conical flasks (250 ml)

(d) Biological pipettes

(e) Colony counter

(f) Water bath

(g) Culture tubes

(h) **Nutrient Agar Medium (NAM):**

Beef extract	3.0 g
Peptone	5.0 g
Agar-agar	15.0 g
Distilled water	1.0 L
pH	6.8±0.2
(after sterilization)	

Preparation of serial dilutions:

5 or 6 culture tubes each containing 9 ml distilled water, were sterilized after plugging. The sample was vigorously

shaken and 1 ml was transferred to the first culture tube. Now 1 ml water from first tube was again transferred to 2nd test tube and so on. Separate pipettes were used for each transfer. The dilutions were 1:10, 1:100, 1:1000, 1:10000, and so on.

Procedure

3 pairs of borosil petri dishes were sterilized in an oven. Now 1 ml of liquid from last 3 dilution was transferred to the petridishes by opening them gently. 15-20 ml medium was poured into these plates. The temperature of NAM should be about 45°C. Both water sample and medium, were mixed by gently rotating the plate. When medium became solidify the plates were inverted and incubated at 37±2°C temperature for 48 hours. The colonies were counted after 48 hours by colony counter and Colony Forming Unit (CFU)/ml was calculated.

Calculation:

$$\mathbf{CFU/ml} = \frac{\textbf{Colonies counted}}{\textbf{Dilution factor}}$$

21. *Most Probable Number (MPN)*

The bacteriological quality of water is based on testing for bacterial species of known excretal origin, particularly organisms of the coliform group. The coliform group of bacteria included all the aerobic and facultatively anaerobic, Gram negative, non-sporulating bacilli, that produce acid and gas from the fermentation of lactose.

Coliform bacteria as typified by *E.coli* and fecal Streptococci residing in the intestinal tract of humans are excreted in feces of humans and other warm blooded animals averaging about 50 million coliforms per gram (Hammer, 1986).

The multiple tube fermentation technique can be used to enumerate positive presumptive and confirmed coliform tests. Result of the test are expressed as the most probable number (MPN), since the count is based on statistical analysis of sets of tube in a series of serial dilution. MPN is by definition is related to a sample volume of 100 ml.

Apparatus and Material

(*a*) Fermentation tubes (50 ml)

(*b*) Durham's tubes

(*c*) Biological pipettes

(*d*) Inoculation loop

(*e*) Incubator

(*f*) Conical flasks

(*g*) Autoclave

(*h*) Spirit lamp

(*i*) Water bath

A. Presumptive Test:

MacConkey's Broth is required for this test. The composition of broth is as given below.

Bile salt or Sodium taurocholate	5.0 g
Peptone	20.0 g
Lactose	10.0 g
NaCl	5.0 g
1% Bromo cresol purple	1.0 ml
Distilled water	1.0 L
Final pH	7.1±0.2
(after sterilization)	

Procedure

Several dilution of the sample for Ganga water prepared. Dilutions above than 1:1000 are required. 5 fermentation tubes for each dilution, were filled with 9 ml of broth and one Durham's vial was put in each tube in inverted condition. Now all the tubes were plugged with cotton and sterilized in autoclave. All the dilutions were shaked vigorously before moving the sample to fermentation tubes. Separate sterilized pipettes were used for different dilutions as well as samples.

Fermentation tubes were gently shaked. Now these tubes were incubated at 35.37°C temperature for 48 hours. All the tubes were examined after 48 hours for gas production.

The tubes, showing gas production in Durham's tube, were recorded as positive (+ve). Even a tiny bubble should be recorded as positive. The tubes showing positive results were now subjected to the confirmatory test. Negative tubes were incubated for further 24 hours to confirm the test.

B. Confirmatory Test

For the **Brilliant Green Lactose Bile (BGLB)** broth is required. This inhibits the growth of lactose fermenters other than coliform; thus, gas formation in BGLB broth constitutes a confirmed test i.e. coliform present. The composition of BGLB broth is given below.

Peptone	10.0 g
Lactose	10.0 g
Oxgall	20.0 g
Brilliant green	0.0133 g
Distilled water	1.0 L
pH	7.2±0.2
(after sterilization)	

Procedure

Fermentation tubes with 10 ml BGLB broth and inverted Durham's tube were prepared. The numbers of tubes should be equal to all +ve tube in the presumptive test. Positive tubes were shook gently. One loopful of liquid was transferred to the tubes having BGLB with the help of inoculation loop. These tubes were incubated at 35-37°C for 48 hours.

The tubes were recorded as positive, which showed the gas production. The number of positive tubes was put in MPN table and then in the formula to calculated the MPN/100 ml.

Calculation:

$$\text{MPN/100 ml} = \frac{\text{MPN table value x 10}}{\text{Starting value}}$$

22. *Plankton*

Collection of plankton

The plankton of river were collected by fine plankton net. The known quantity of water was pass through filter, and concentrated in the desired quantity. 1 ml of the concentrate was taken and placed in a Sedgwick Rafter counting cell. Counting of the organisms was done by appling the following formula:

$$\textbf{Plankton/l} = \frac{\textbf{(a x 1000)c}}{\textbf{1}}$$

Where, a = average no. of phytoplankton in one small counting chamber of Sedgwick Rafter counting cell

c = ml. of plankton concentrate

l = Volume of original water filtered in litre.

ISOLATION AND IDENTIFICATION OF BACTERIAL CULTURE:

Isolation of Bacteria:

Two methods were use to isolating bacteria are (i) pour plate and (ii) streak plate method.

(i) **Pour Plate Method:** In this method 5 or 6 culture tubes each containing 9 ml distilled water, were sterilized after plugging for serial dilution. The sample was vigorously shaken and 1 ml was transferred to the first culture tube. Now 1ml water from first tube was again transferred to 2nd test tube and so on. Separate pipettes were used for each transfer. The dilutions were 1:10, 1:100, 1:1000, 1:10000, and so on.

3 pairs of borosil petridishes were sterilized in an oven. Now 1 ml of liquid from last 3 dilution was transferred to the petridishes by opening them gently. 15-20 ml medium was poured into these plates. The temperature of NAM should be about 45°C. Both water sample and medium, were mixed by gently rotating the plate. When medium became solidify the plates were inverted and incubated at 37±2°C temperature for 48 hours.

(ii) **Streak Method:** From these poured plates, individual colony was transferred with the help of sterilized loop (by incineration), on another media plate and streaked. These plates were incubated for further 48 hours at 37±2°C, for identification of pure colony.

Identification of Bacteria:

(i) **Incubation Temperature:** Each isolated colony was incubated at, below 20°C (Psycrophileic), 20-45°C (Mesophileic), and 45°C and above (Thermophileic).

(ii) **Gram Staining:** Individual isolated colony were stained by Gram stain and differentiated in Gram positive and Gram negative.

Principle

Gram staining is a method of differential staining which employs two stains to distinguish usually morphologically similar but different species of bacteria. The technique involves staining of bacteria by crystal violet (primary stain) and safranin or basic fuchsin (counter stain). The bacteria retaining the primary stain are called gram-positive, while the bacteria yielding out the primary stain by alcohol washing and retaining the counter stain are called gram-negative.

Reagents

A. *Crystal Violet*

(a) Dissolve 2 g of crystal violet in 20 ml of 95% ethyl alcohol.

(b) Dissolve 0.8 g of ammonium oxalate in 80 ml. of distilled water. Mix the two solutions (a and b).

B. *Gram's Iodine*

Dissolve 1 g of iodine and 2 g of KI in 300 ml of distilled water.

C. *Ethyl alcohol,* 95% or Acetone alcohol (1:1).

D. *Sarfanin,* 0.5%.

Procedure

1. Prepare a suspension of bacterial colony in sterile 0.85% NaCl (Saline) to disperse the bacterial-cells.

2. Take a slide and mark an area on it by a glass marking pencil.

3. Prepare a smear of the bacterial suspension in the area marked on the glass slide. Air dry and fix the smear by very slight heating.

4. Pour the crystal violet primary stain on it and allow to react for 1-2 minutes. Wash the slide with water to remove the excess stain.

5. Put Gram's iodine solution on the smear and keep for about a minute.

6. Wash the slide with 95% alcohol or acetone alcohol till the last traces of colour come out.

7. Wash the slide with water and stain with safranin for about 30 sec.-1 minute.

8. Give the final washing with water, air dry and observe under a high power microscope preferably with an oil immersion lens.

9. The bacteria showing violet colour are gram-positive, while the bacteria showing red or pink colour and gram-negative.

(iii) **Aerobic, Facultative anerobic or Anaerobic**: Individual culture was inoculated in sterilized liquid Soyacasien Digest medium (Hi-Media) and by the growth pattern, differentiation was stablished wheather they are Aerobic, Facultative anerobic or Anaerobic cultures.

(iv) **Motility:** Culture were mounted on groved slide and observed under light microscope, motality of the microorganism.

(v) **Fermentation:** Isolated bacteria was further check for its fermenting nature, i.e., growth with gas and acid production in MacConkeys Broth medium.

Procedure

MacConkey's Broth is required for this test. The composition of broth is as given below.

Bile salt or Sodium taurocholate	5.0 g
Peptone	20.0 g
Lactose	10.0 g
NaCl	5.0 g
1% Bromo cresol purple	1.0 ml
Distilled water	1.0 L
Final pH (after sterilization)	7.1±0.2

Several dilution of the sample for Ganga water prepared. Dilutions above than 1:1000 are required. 5 fermentation tubes for each dilution, were filled with 9 ml of broth and one Durham's vial was put in each tube in inverted condition. Now all the tubes were plugged with cotton and sterilized in autoclave. All the dilutions were shaked vigorously before moving the sample to fermentation tubes. Separate sterilized pipettes were used for different dilutions as well as samples. Fermentation tubes were gently shaked. Now these tubes were incubated at 35.37°C temperature for 48 hours. All the tubes were examined after 48 hours for gas production.

The tubes, showing gas production in Durham's tube, were recorded as positive (+ve). Even a tiny bubble should be recorded as positive. The tubes not showing gas or acid production are noted as negative (-ve).

Statistical Analysis of Data:

The data for all physico-chemical and microbiological parameters were collected and the data were analysed for calculating mean, standard deviation of each parameter and correlation as per Goel (1999) with the help of computer on Microsoft Excel.

1. $$\textbf{Mean } (\bar{x}) = \frac{\sum X}{N}$$

Where, X = Individual reading of parameter.

ΣX = Total of all readings for parameter.

N = Number of samples.

2. ***Standard Deviation (SD or σ)***

$$SD = \sqrt{\frac{\sum (x - \bar{x})^2}{n - 1}}$$

Where, X = Individual reading of parameter

$\bar{x}$ = Mean of ΣX

n = Number of samples.

3. ***Coefficient of Correlation (r)***

$$r = \frac{\sum (x - \bar{x})(y - \bar{y})}{\sqrt{\sum (x - \bar{x})^2 \sum (y - \bar{y})^2}}$$

Where, X = Individual reading of 1st parameter

$\bar{x}$ = Mean of ΣX

Y = Individual reading of 2nd parameter

$\bar{y}$ = Mean of ΣY

4

Observations

A. PHYSICO-CHEMICAL PARAMETERS

Quality of any water is basically depends on its physico-chemical, biological and microbiological parameters. The physico-chemical of the river Ganga studied at Hardwar were water temperature (°C), turbidity (JTU), conductivity (μmhos/cm^2), velocity (m/sec), total solids (mg/l), total dissolved solids (mg/l), total suspended solids (mg/l), pH, free carbon dioxide (mg/l), dissolved oxygen (mg/l), BOD (mg/l), COD (mg/l), Total alkalinity (mg/l), chloride (mg/l), total phosphate (mg/l), total organic carbon (mg/l), hardness (mg/l), ions – (a) sodium (mg/l), (b) potassium (mg/l), (c) sulphate (mg/l), (d) nitrate (mg/l), and heavy metals – (a) lead (mg/l), (b) copper (mg/l), (c) chromium (mg/l), (d) cadmium (mg/l), (e) zinc (mg/l). The observations made during April 1998 to March 1999 and April 1999 to March 2000 are presented in the Table 4A.1 to 4A.8, there interrelationship in Fig. 4A.1 to 4A. 40 and linear graph for ions and heavy metals are shown in Figs. 4A.41-4A.49.

Sampling Station A (Har-ki-Pauri): The physical parameters of the sampling station A are presents in Table 4A.1 and Figs. 4A.1 to 4A.3 and the chemical parameters are shown in the Tables 4A.1-4A.2 and Figs. 4A.4-4.A10.

1. Water Temperature

The water temperature of river Ganga at this sampling station during total study period was observed to be lowest in winter season (1999-2000), 8.80°C ± 0.63, which goes highest up

Table 4A.1. Seasonal Variation in Physico-chemical Parameter of River Ganga at Sampling Station A (1998-2000)

Physico-chemical Parameters	1998 to 1999				1999 to 2000			
	Summer	Monsoon	Winter	Average	Summer	Monsoon	Winter	Average
1. Water Temperature (°C)	18.00 ± 0.71	19.00 ± 1.56	10.39 ± 0.60	15.80 ± 4.71	19.44 ± 0.46	18.81 ± 0.24	8.80 ± 0.63	15.69 ± 5.97
2. Turbidity (JTU)	32.22 ± 3.84	541.28 ± 15.98	19.33 ± 1.50	197.61 ± 297.69	33.50 ± 2.18	575.22 ± 19.02	20.78 ± 1.99	209.83 ± 316.50
3. Conductivity (μmhos/cm^2)	138.00 ± 1.00	337.56 ± 7.54	99.56 ± 2.46	191.70 ± 127.77	137.33 ± 1.41	364.33 ± 31.65	111.89 ± 4.54	204.52 ± 138.99
4. Velocity (m/sec)	1.59 ± 0.28	1.88 ± 0.09	0.73 ± 0.05	1.40 ± 0.60	1.84 ± 0.07	1.99 ± 0.01	0.92 ± 0.06	1.58 ± 0.58
5. Total Solids (mg/l)	552.70 ± 47.92	3826.88 ± 178.39	158.00 ± 2.92	1512.53 ± 2013.98	631.78 ± 25.64	3887.64 ± 278.74	172.11 ± 16.71	1563.84 ± 2025.55
6. Total Dissolved Solids (mg/l)	220.13 ± 6.94	850.56 ± 22.32	48.33 ± 4.50	373.01 ± 422.40	279.00 ± 43.91	997.67 ± 77.31	33.33 ± 8.93	436.67 ± 501.13
7. Total Suspended Solids (mg/l)	332.57 ± 45.63	2976.62 ± 178.57	108.11 ± 4.23	1139.00 ± 1595.12	440.00 ± 60.92	2889.97 ± 244.38	138.78 ± 15.91	1156.25 ± 1508.98
8. pH	7.30 ± 0.27	7.57 ± 0.23	7.07 ± 0.10	7.32 ± 0.25	7.29 ± 0.26	7.19 ± 0.14	7.03 ± 0.20	7.17 ± 0.13
9. Free Carbon Dioxide (mg/l)	2.21 ± 0.29	4.97 ± 0.20	1.75 ± 0.07	2.98 ± 1.74	2.15 ± 0.13	4.77 ± 0.40	1.60 ± 0.07	2.84 ± 1.69
10. Dissolved Oxygen (mg/l)	9.52 ± 0.89	7.20 ± 0.23	11.74 ± 0.43	9.49 ± 2.27	10.38 ± 0.99	7.29 ± 0.24	10.87 ± 0.47	9.51 ± 1.94
11. B.O.D. (mg/l)	2.36 ± 0.27	3.02 ± 0.56	1.38 ± 0.10	2.25 ± 0.83	2.30 ± 0.12	2.73 ± 0.47	1.57 ± 0.51	2.20 ± 0.59

Table 4A.1. (Contd.)

12. C.O.D. (mg/l)	9.09 ± 1.02	5.31 ± 2.15	11.62 ± 0.40	8.67 ± 3.18	8.86 ± 0.46	6.05 ± 1.82	10.52 ± 1.97	8.48 ± 2.26
13 Total Alkalinity (mg/l)	66.67 ± 2.50	48.22 ± 0.67	35.33 ± 0.50	50.07 ± 15.75	67.89 ± 1.62	46.44 ± 1.33	37.89 ± 1.17	50.74 ± 15.45
14 Chloride (mg/l)	2.96 ± 0.03	11.55 ± 0.28	1.97 ± 0.04	5.49 ± 5.27	2.92 ± 0.05	11.32 ± 0.38	1.72 ± 0.23	5.32 ± 5.23
15. Total Phosphate (mg/l)	0.02 ± 0.00	0.11 ± 0.01	0.07 ± 0.01	0.07 ± 0.04	0.03 ± 0.00	0.13 ± 0.01	0.07 ± 0.00	0.07 ± 0.05
16. Total Organic Carbon (mg/l)	3.37 ± 0.27	5.25 ± 0.15	4.14 ± 0.03	4.25 ± 0.94	3.41 ± 0.15	5.35 ± 0.20	4.09 ± 0.09	4.28 ± 0.98
17 Hardness (mg/l)	83.78 ± 1.30	105.53 ± 2.21	96.00 ± 0.71	95.10 ± 10.91	99.11 ± 5.69	111.73 ± 4.71	90.11 ± 7.47	100.32 ± 10.86
18. Ions								
(*a*) Sodium (mg/l)	9.3702 ± 0.02	23.6357 ± 0.03	11.4796 ± 0.02	14.8285 ± 7.70	10.1795 ± 0.05	26.6435 ± 0.02	13.3803 ± 0.05	16.7344 ± 8.73
(*b*) Potassium (mg/l)	1.2230 ± 0.03	2.5791 ± 0.03	1.6225 ± 0.02	1.8082 ± 0.70	1.2230 ± 0.02	2.6151 ± 0.02	1.6310 ± 0.02	1.8230 ± 0.72
(*c*) Sulphate (mg/l)	18.3000 ± 0.03	24.1000 ± 0.03	19.5000 ± 0.05	20.6333 ± 3.06	18.9900 ± 0.02	24.1000 ± 0.02	18.2200 ± 0.04	20.4367 ± 3.20
(*d*) Nitrate (mg/l)	0.0140 ± 0.01	0.0600 ± 0.01	0.0300 ± 0.02	0.0347 ± 0.02	0.0140 ± 0.01	0.0600 ± 0.01	0.0200 ± 0.01	0.0313 ± 0.03

± = Standard Deviation.

Table 4A.2. Seasonal Variation in Heavy Metals of River Ganga at Sampling Station A (1998-2000)

Seasons Parameters	1998 to 1999				1999 to 2000			
	Summer	Monsoon	Winter	Average	Summer	Monsoon	Winter	Average
Lead [Pb] (mg/l)	0.0773 ± 0.0031	0.0767 ± 0.0015	0.0630 ± 0.0026	0.0723 ± 0.0081	0.0773 ± 0.0021	0.0788 ± 0.0011	0.0641 ± 0.0029	0.0734 ± 0.0081
Copper [Cu] (mg/l)	0.0175 ± 0.0006	0.0185 ± 0 0005	0.0164 ± 0.0002	0.0174 ± 0.0011	0.0190 ± 0.0013	0.0183 ± 0.0005	0.0170 ± 0.0002	0.0181 ± 0.0010
Chromium [Cr] (mg/l)	0.0303 ± 0.0021	0.0326 ± 0.0018	0.0301 ± 0.0004	0.0310 ± 0.0014	0.0346 ± 0.0003	0.0340 ± 0.0017	0.0270 ± 0.0005	0.0319 ± 0.0042
Cadmium [Cd] (mg/l)	0.0028 ± 0.0006	0.0031 ± 0.0003	0.0029 ± 0.0001	0.0029 ± 0.0001	0.0034 ± 0.0003	0.0037 ± 0.0002	0.0026 ± 0.0002	0.0032 ± 0.0006
Zinc [Zn] (mg/l)	3.2067 ± 0.0513	3.2416 ± 0.0177	3.1783 ± 0.0516	3.2088 ± 0.0317	3.2503 ± 0.0603	3.2469 ± 0.0140	3.2010 ± 0.0010	3.2328 ± 0.0275

± = Standard Deviation.

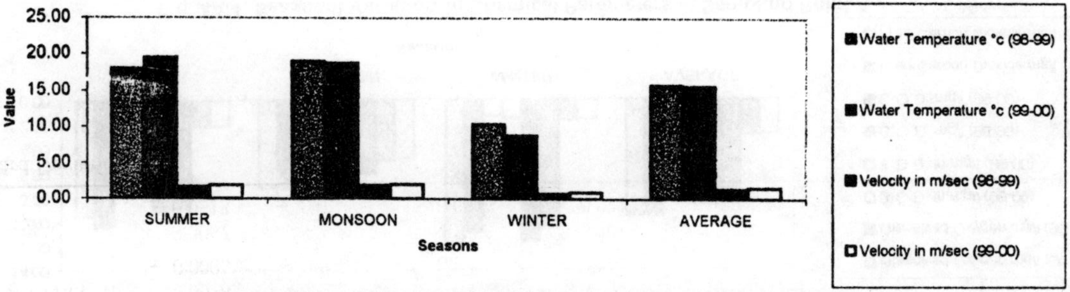

Fig. 4A.1. Seasonal Variation in Physical Parameters at Sampling Station A (1998-2000)

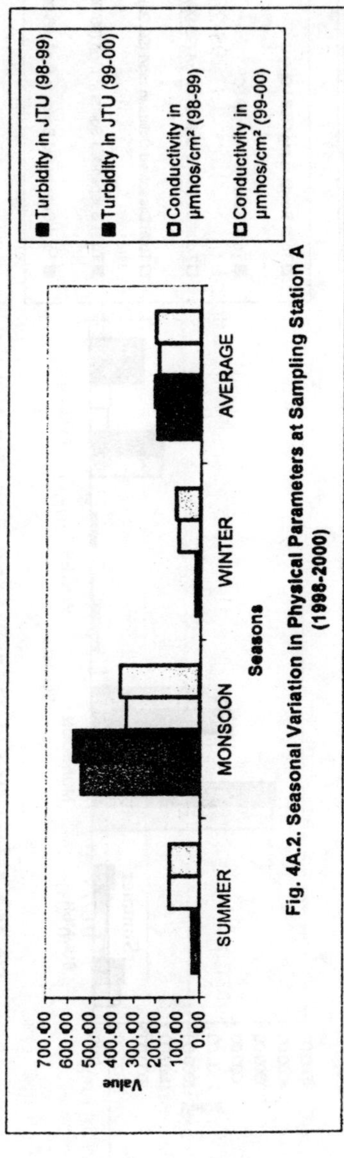

Fig. 4A.2. Seasonal Variation in Physical Parameters at Sampling Station A (1998-2000)

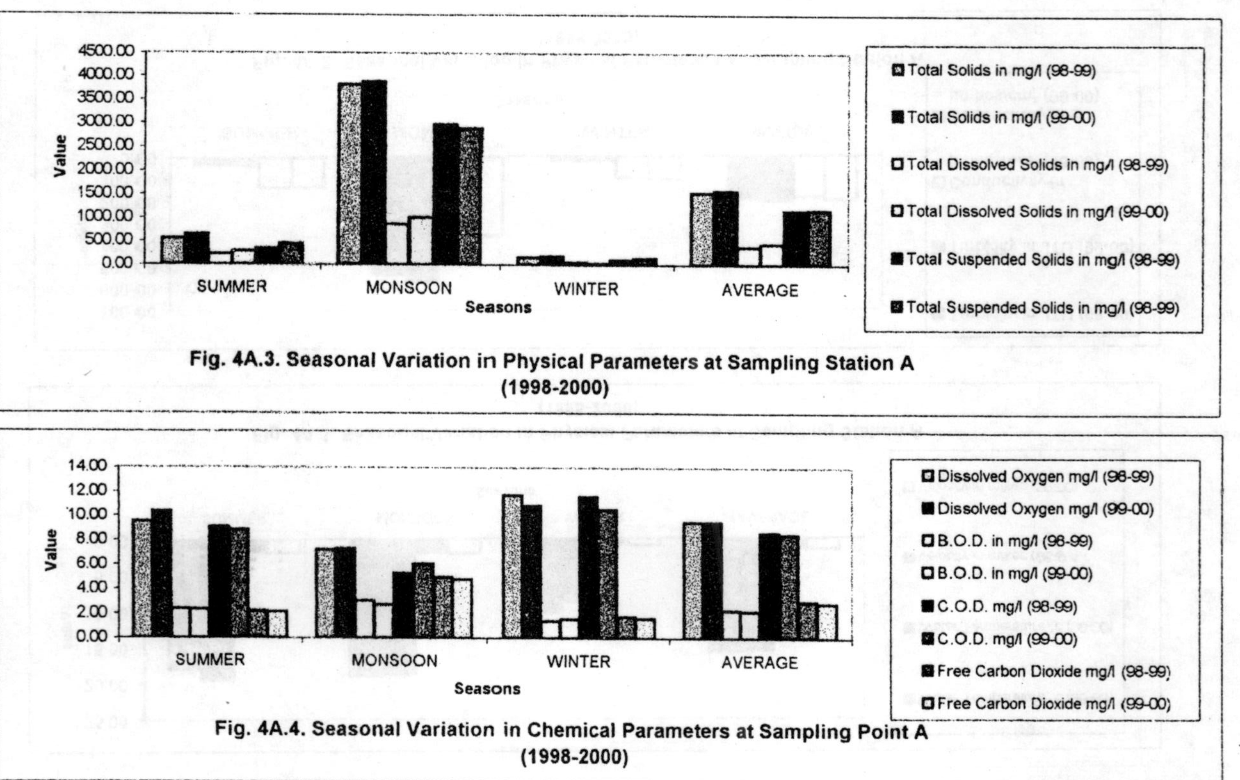

Fig. 4A.3. Seasonal Variation in Physical Parameters at Sampling Station A (1998-2000)

Fig. 4A.4. Seasonal Variation in Chemical Parameters at Sampling Point A (1998-2000)

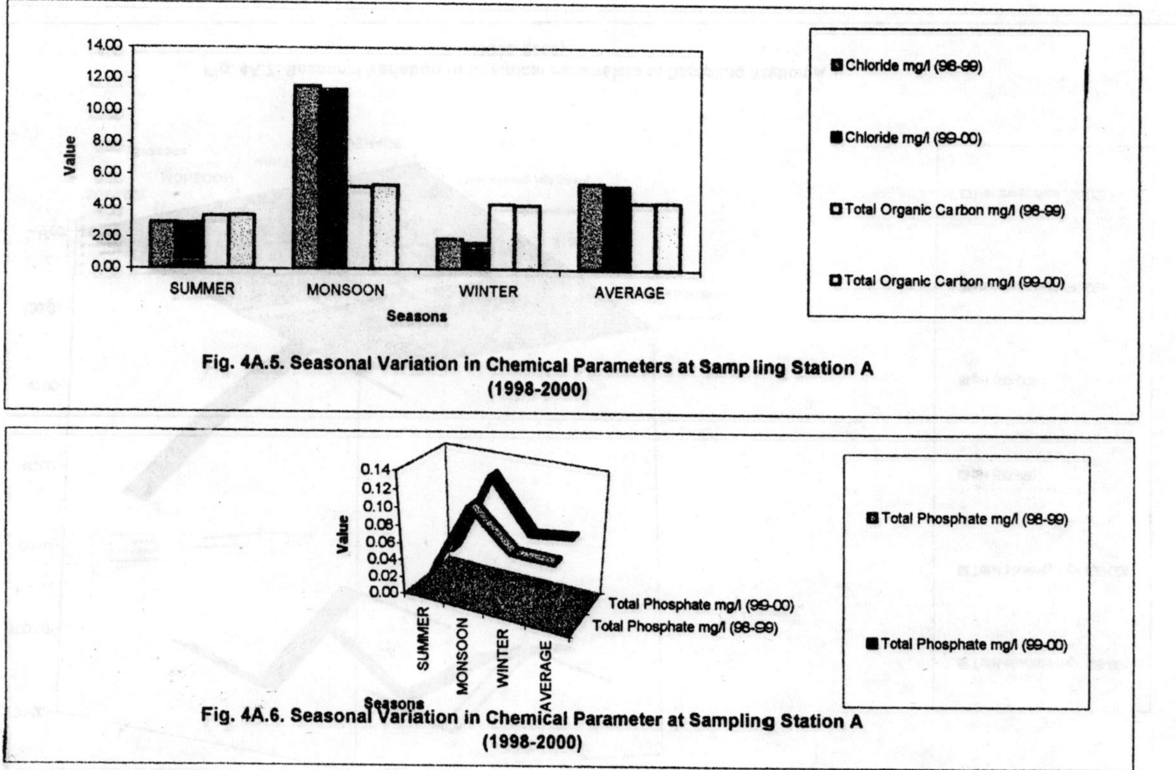

Fig. 4A.5. Seasonal Variation in Chemical Parameters at Samp ling Station A (1998-2000)

Fig. 4A.6. Seasonal Variation in Chemical Parameter at Sampling Station A (1998-2000)

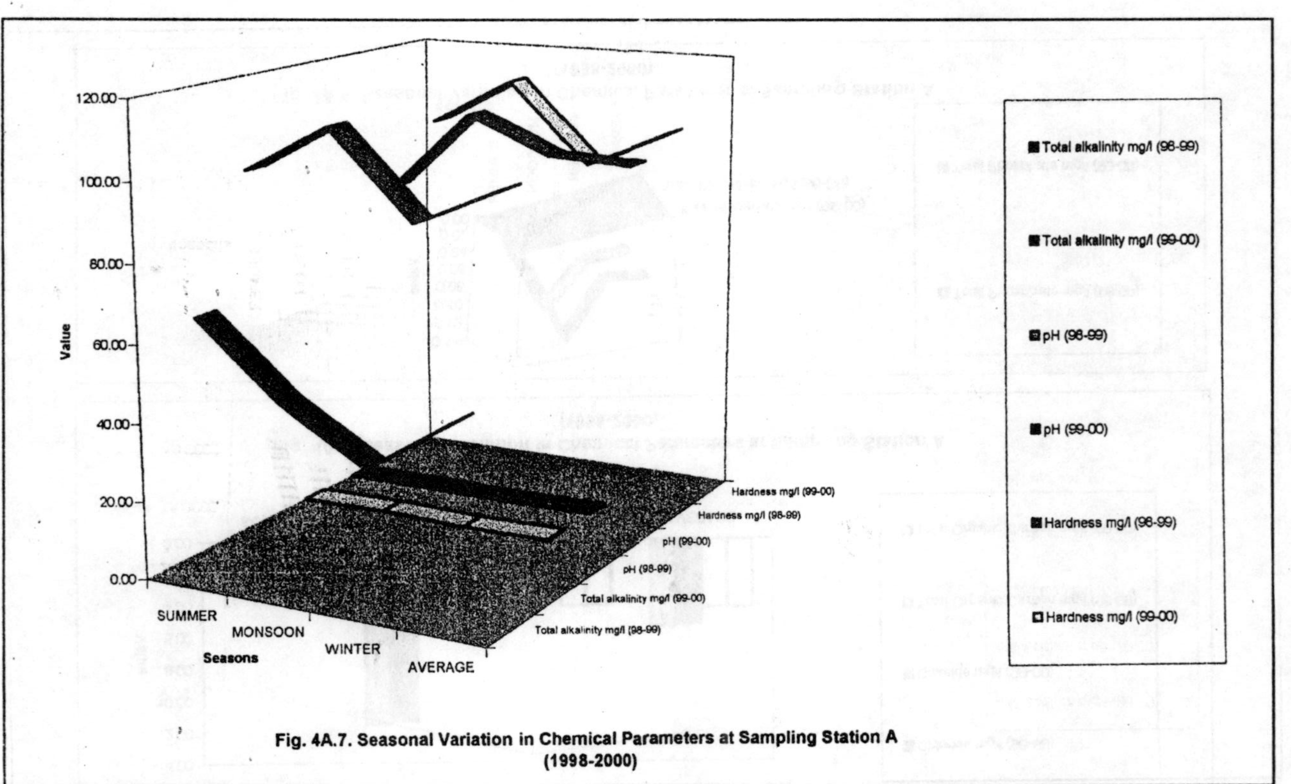

Fig. 4A.7. Seasonal Variation in Chemical Parameters at Sampling Station A (1998-2000)

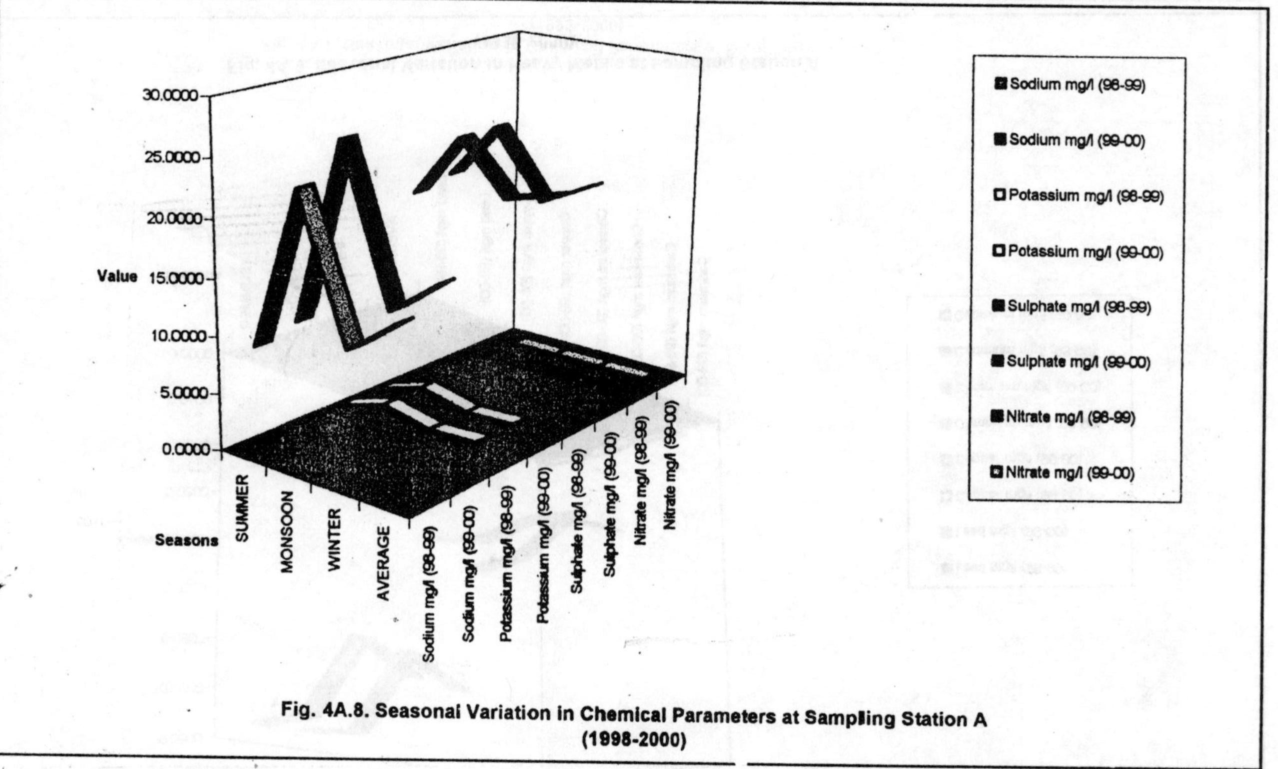

Fig. 4A.8. Seasonal Variation in Chemical Parameters at Sampling Station A (1998-2000)

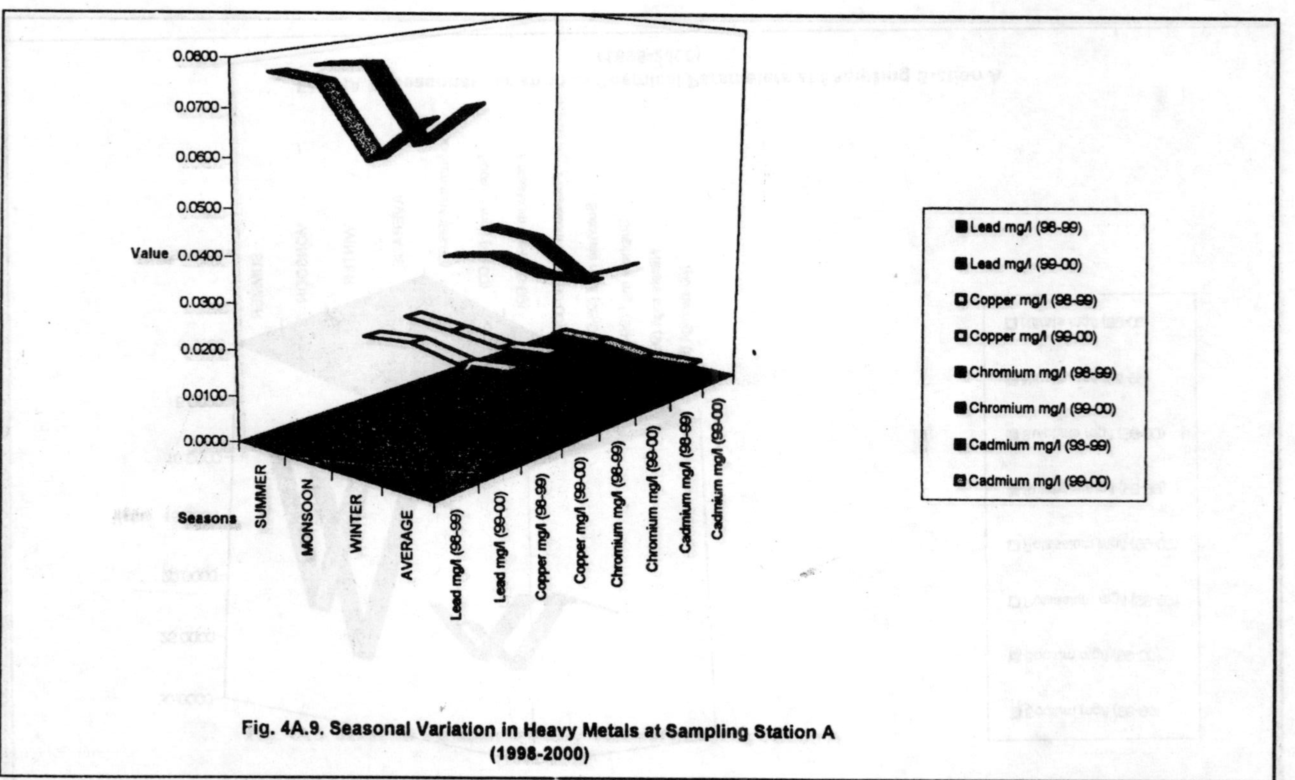

Fig. 4A.9. Seasonal Variation in Heavy Metals at Sampling Station A (1998-2000)

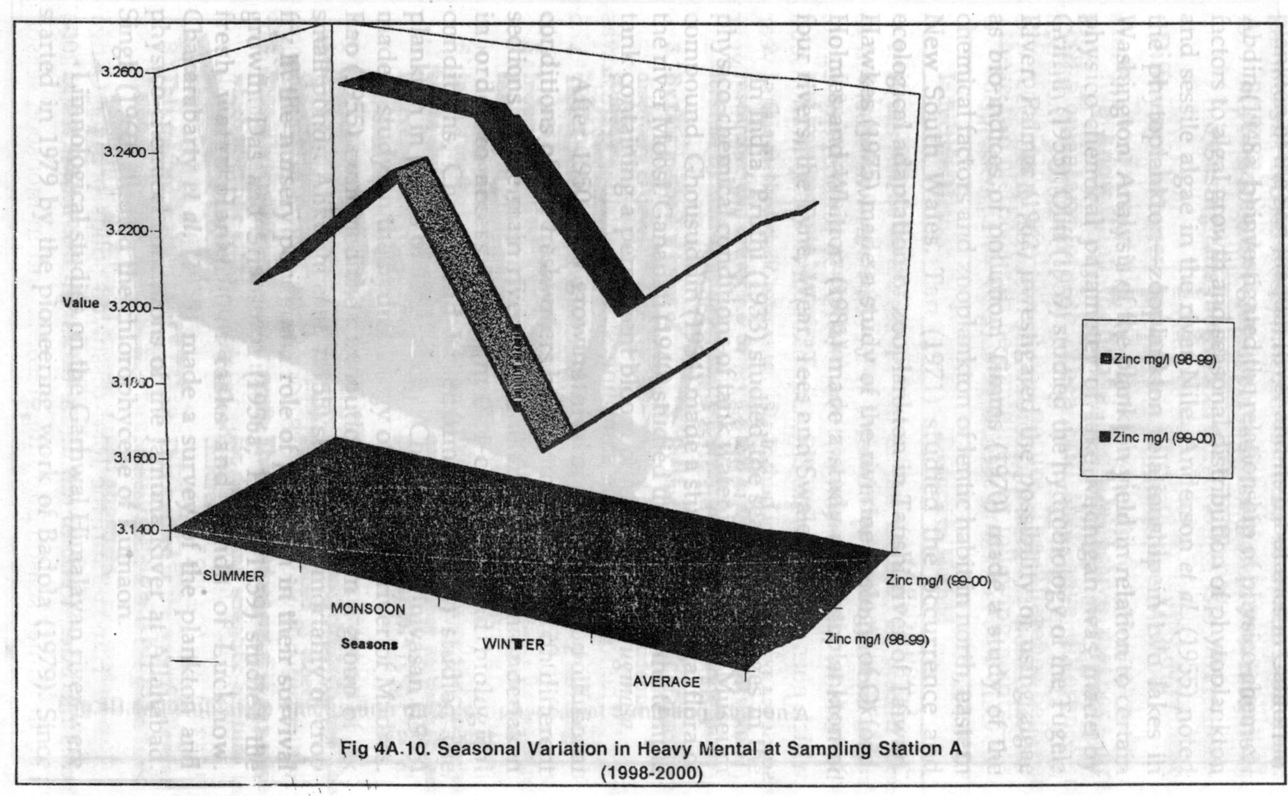

Fig 4A.10. Seasonal Variation in Heavy Mental at Sampling Station A (1998-2000)

to 19.44°C ±0.46 in summer season (1999-2000). The annual average recorded for first year (1998-1999) 15.80°C ±4.71 and for second year (1999-2000) it was calculated 15.69°C ±5.97.

2. Turbidity

The turbidity of water at this spot observed to be minimum in 1998-1999's winter season 19.33 JTU ±1.50, which goes maximum up to 575.22 JTU ±19.02 in 1999-2000's monsoon season. The annual mean recorded for first year (1998-1999) was 197.61 JTU ±297.69 and second year (1999-2000) was 209.83 JTU ±316.50.

3. Conductivity

The conductivity of water at Har-ki-Pauri was observed to be least in winter season of 1998-1999, 99.56 μmhos/cm^2 ±2.46, which goes highest up to 364.33 μmhos/cm^2 ±31.65 in monsoon season of 1999-2000. The annual average observed for first year (1998-1999) was 191.70 μmhos/cm^2 ±127.77 and second year (1999-2000) was 204.50 μmhos/cm^2 ±138.99.

4. Velocity

The velocity of this sampling station at river Ganga was noted to be minimum in winter season of 1998-1999, 0.73 m/sec ±0.05, which goes maximum up to 1.99 m/sec ±0.01 in monsoon season of 1999-2000. The annual average calculated for first year (1998-1999) was 1.40 m/sec ±0.60 and second year (1999-2000) was 1.58 m/sec ±0.58.

5. Total Solids

The total solids of river Ganga at this spot was recorded to be lowest in 1998-1999's winter season, 158.0 mg/l ±2.92, which goes highest up to 3887.64 mg/l ±278.74 in 1999-2000's monsoon season. The annual mean recorded for first year (1998-1999), 1512.53 mg/l ±2013.98 and second year (1999-2000) was 1563.84 mg/l ±2025.55.

6. Total Dissolve Solids

At Har-ki-Pauri the total dissolved solids of river Ganga was recorded to be minimum and maximum both in 1999-2000 i.e., in winter season 33.33 mg/l ±8.93, and in monsoon season

997.67 mg/l ±77.31 respectively. The annual average recorded for first year (1998-1999) 373.01 mg/l ±422.40 and second year (1999-2000) 436.67 mg/l ±501.13.

7. Total Suspended Solids

The total suspended solids of river Ganga at this sampling station were observed to be lowest and highest in (1998-1999). Lowest was in winter season 108.11 mg/l ±4.23, which goes highest up to 2976.30 mg/l ±178.57 in monsoon season of 1998-1999. The annual average recorded for first year (1998-1999) was 1139.0 mg/l ±1595.12 and second year (1999-2000) 1156.25 mg/l ±1508.98.

8. pH

At this sampling station the pH of river Ganga was noted to be least in winter season (1999-2000) 7.03 ±0.20, which goes highest up to 7.57 ±0.23 in monsoon season (1998-1999). The annual average recorded for first year (1998-1999) was calculated 7.32 ±0.25 and second year (1999-2000) was 7.17 ±0.13.

9. Free Carbon Dioxide

The free carbon dioxide of river Ganga at this sampling station was observed to be lowest in winter season (1999-2000), 1.60 mg/l ±0.07, which goes highest up to 4.97 mg/l ±0.20 in monsoon season (1998-1999). The annual mean recorded for first year (1998-1999) 2.98 mg/l ±1.74 and second year (1999-2000) 2.84 mg/l ±1.69.

10. Dissolved Oxygen

The dissolved oxygen of river Ganga at this sampling station was recorded to be minimum in monsoon season (1998-1999) 7.20 mg/l ±0.23, which goes maximum up to 11.74 mg/l ± 0.43 in winter season (1998-1999). The annual average observed for first year (1998-1999), 9.49 mg/l ±2.27 and second year (1999-2000), was 9.51 mg/l ± 1.94.

11. Biological Oxygen Demand (BOD)

In the same year (1998-1999) the biological oxygen demand (BOD) of river Ganga at this spot was noted to be lowest in winter season 1.38 mg/l ±0.10, which goes highest up to 3.02

mg/l ±0.56 in monsoon season. The annual average recorded for first year (1998-1999) 2.25 mg/l ±0.83 and second year (1999-2000) 2.20 mg/l ±0.59.

12. Chemical Oxygen Demand (COD)

The chemical oxygen demand (COD) of river Ganga at this sampling station was observed to be least in monsoon season (1998-1999) 5.31 mg/l ±2.15, which goes highest up to 11.62 mg/l ±0.40 in winter season (1998-1999). The annual mean calculated for first year (1998-1999). 8.67 mg/l ±3.18 and second year (1999-2000) 8.48 mg/l ±2.26.

13. Total alkalinity

The Total alkalinity of water at Har-ki-Pauri was noted to be minimum in winter season (1998-1999), 35.33 mg/l ±0.50, which goes highest up to 67.89 mg/l ±1.62 in summer season (1999-2000). The annual average recorded for first year (1998-1999) was 50.07 mg/l ±15.75 and second year (1999-2000) was 50.74 mg/l ±15.45.

14. Chloride

The chloride in the water of river Ganga at Har-ki-Pauri this spot was recorded to be lowest in winter season (1999-2000), 1.72 mg/l ±0.23, which goes highest up to 11.55 mg/l ±0.28 in monsoon season (1998-1999). The annual average calculated for first year (1998-1999) was 5.49 mg/l ±5.27 and second year (1999-2000) was 5.32 mg/l ±5.23.

15. Total Phosphate

At this sampling station the total phosphate of river Ganga reveled that minimum was in summer season (1998-1999), 0.02 mg/l ±0.00, which goes maximum up to 0.13 mg/l ±0.01 in monsoon season (1999-2000). The annual average observed for first year (1998-1999 was 0.07 mg/l ±0.04 and second year (1999-2000) was 0.07 mg/l ±0.05.

16. Total Organic Carbon (TOC)

The total organic carbon (TOC) at this spot of river Ganga was recorded to be lowest in summer season (1998-1999), 3.37 mg/l ±0.27, which goes maximum up to 5.35 mg/l ±0.20

in monsoon season (1999-2000). The annual mean noted for first year (1998-1999) was 4.25 mg/l ±0.94 and second year (1999-2000) was 4.28 mg/l ±0.98.

17. Hardness

The hardness at this sampling station of river Ganga was observed to be least in summer season (1998-1999), 83.78 mg/l ±0.30, which goes maximum up to 111.73 mg/l ±4.71 in monsoon season (1999-2000). This annual average observed for first year (1998-1999) was 95.10 mg/l ±10.91 and second year (1999-2000) was 100.32 mg/l ±10.86.

18. Ions

(a) Sodium

The sodium at this spot of river Ganga was recorded to be minimum in summer season (1998-1999), 9.3702 mg/l ±0.02, which goes maximum up to 26.6435 mg/l± 0.02 in monsoon season (1999-2000). The annual average was found for first year (1998-1999) 14.82585 mg/l ±7.70 and was 16.7344 mg/l ±8.73 for second year (1999-2000).

(b) Potassium

At this sampling station the potassium of river Ganga was observed to be lowest in summer season (1998-1999), 1.2230 mg/l ±0.03, which goes highest up to 2.6151 mg/l ±0.02 in monsoon season (1999-2000). The annual mean calculated for first year (1998-1999) was 1.8082 mg/l ±0.70 and second year (1999-2000) was 1.8230 mg/l ±0.72.

(c) Sulphate

At this sampling station the sulphate of river Ganga was recorded to be minimum in winter season (1999-2000), 18.2200 mg/l ±0.04, which goes maximum up to 24.1000 mg/l ±0.02 in monsoon season (1999-2000). The annual average was found 20.6333 mg/l ±3.06 for first year (1998-1999) and 20.4367 mg/l ±3.20 for second year (1999-2000).

(d) Nitrate

The nitrate in water at Har-ki-Pauri was observed to be lowest in summer season (1998-1999), 0.0140 mg/l ±0.01, which goes highest up to 0.0600 mg/l ±0.01 in monsoon season (1999-2000). The annual mean recorded for first year (1998-1999) was 0.0347 mg/l ±0.02 and second year (1999-2000) was 0.0313 mg/l ±0.03.

19. Heavy Metals

(a) Lead (Pd)

The lead at this sampling station of river Ganga was recorded to be minimum in winter season (1998-1999), 0.0630 mg/l ±0.0026, which goes maximum up to 0.0788 mg/l ±0.0011 in monsoon season (1999-2000). The annual average calculated for first year (1998-1999) was 0.0723 mg/l ±0.0081 and second year (1999-2000) was 0.0734 mg/l ±0.0081.

(b) Copper (Cu)

At this spot the copper was recorded to be lowest in winter season (1998-1999), 0.0164 mg/l ±0.0002, which goes highest up to 0.0185 mg/l ±0.0005 in monsoon season (1998-1999). The annual average observed for first year (1998-1999) was 0.0174 mg/l ±0.0011 and second year (1999-2000) was 0.0181 mg/l ±0.0010.

(c) Chromium (Cr)

At this sampling station the chromium of river Ganga was recorded to be least in winter season (1999-2000), 0.0270 mg/l ±0.0005, which goes optimum up to 0.0346 mg/l ±0.0003 in summer season (1999-2000). The annual average calculated for first year (1998-1999) was 0.0310 mg/l ±0.0014 and second year (1999-2000) was 0.0319 mg/l ±0.0042.

(d) Cadmium (Cd)

The cadmium at this sampling station of river Ganga was observed to be lowest in winter season (1999-2000), 0.0026 mg/l ±0.0002, which goes highest up to 0.0037 mg/l ±0.0002 in monsoon season (1999-2000). The annual average observed for first year (1998-1999) was 0.0029 mg/l ±0.0001 and second year (1999-2000) was 0.0032 mg/l ±0.0006.

(e) Zinc (Zn)

At this spot the zinc was recorded to be minimum in winter season (1998-1999), 3.1783 mg/l ±0.0516, which goes maximum up to 3.2503 mg/l ±0.0603 in summer season (1999-2000). The annual mean calculated for first year (1998-1999) was 3.2088 mg/l ±0.0317 and second year (1999-2000) was 3.2328 mg/l ±0.0275.

SAMPLING STATION B (Mayapuri Ghat)

Physico-chemical Parameters: The Physical parameters of the sampling station B are presents in Table 4A.3 and Figs. 4A.11 to 4A.13 and the Chemical parameters are in Table 4A.3-4A.4 and Figs. 4A.14-4A.20.

1. Water Temperature

The water temperature of river Ganga at the sampling station during total study period was observed to be lowest in winter season (1999-2000), 9.11°C ±0.49, which goes highest up to 19.44°C ±0.30 in monsoon season (1999-2000). The annual average recorded for first year (1998-1999) 15.67°C ±4.32 and for second year (1999-2000) it was calculated 15.98°C ±5.95.

2. Turbidity

The turbidity of water at this spot observed to be minimum in 1998-1999's winter season 19.67 JTU ±1.00, which goes maximum up to 580.67 JTU ±28.78 in 1999-2000's monsoon season. The annual mean recorded for first year (1998-1999) was 199.93JTU ±301.79 and second year (1999-2000) was 212.19 JTU ±319.19.

3. Conductivity

The conductivity of water at Har-ki-Pauri was observed to be least in winter season of 1998-1999, 101.33 µmhos/cm^2 ±2.06, which goes highest up to 344.89 µmhos/cm^2 ±19.27 in monsoon season of 1999-2000. The annual average observed for first year (1998-1999) was 193.41 µmhos/cm^2 ±130.35 and second year (1999-2000) was 198.52 µmhos/cm^2 ±127.47.

4. Velocity

The velocity of this sampling station at river Ganga was noted to me minimum in winter season of 1998-1999, 0.77 m/sec ±0.01, which goes maximum up to 1.91 m/sec ±0.07 in summer season of 1999-2000. The annual average calculated for first year (1998-1999) was 1.28 m/sec ±0.58 and second year (1999-2000) was 1.57 m/sec ±0.57.

Table 4A.3. Seasonal Variation in Physico-chemical Parameter of River Ganga at Sampling Station B (1998-2000)

Physico-chemical	1998 to 1999				1999 to 2000			
Parameters	Summer	Monsoon	Winter	Average	Summer	Monsoon	Winter	Average
1. Water Temperature (°C)	17.56 ± 0.73	18.72 ± 0.79	10.72 ± 0.71	15.67 ± 4.32	19.39 ± 0.49	19.44 ± 0.30	9.11 ± 0.49	15.98 ± 5.95
2. Turbidity (JTU)	31.78 ± 1.86	548.33 ± 20.83	19.67 ± 1.00	199.93 ± 301.79	35.11 ± 2.52	580.67 ± 28.78	20.78 ± 4.54	212.19 ± 319.19
3. Conductivity (μmhos/cm²)	136.33 ± 1.22	342.56 ± 12.90	101.33 ± 2.06	193.41 ± 130.35	138.78 ± 2.17	344.89 ± 19.27	111.89 ± 4.54	198.52 ± 127.47
4. Velocity (m/sec)	1.16 ± 0.02	1.91 ± 0.10	0.77 ± 0.01	1.28 ± 0.58	1.91 ± 0.07	1.88 ± 0.04	0.92 ± 0.06	1.57 ± 0.57
5. Total Solids (mg/l)	536.86 ± 25.74	3847.11 ± 193.48	156.44 ± 1.94	1513.47 ± 2029.92	596.00 ± 50.97	4117.00 ± 250.75	164.33 ± 12.32	1625.78 ± 2168.23
6. Total Dissolved Solids (mg/l)	223.49 ± 13.31	863.22 ± 35.81	48.56 ± 4.95	378.42 ± 428.86	328.22 ± 57.62	1090.33 ± 98.25	35.22 ± 12.26	484.59 ± 544.66
7. Total Suspended Solids (mg/l)	313.37 ± 24.69	2983.89 ± 200.26	108.22 ± 3.70	1135.16 ± 1604.33	270.44 ± 49.62	3026.67 ± 210.41	129.11 ± 17.35	1142.07 ± 1633.63
8. pH	7.19 ± 0.19	7.71 ± 0.14	7.08 ± 0.09	7.33 ± 0.34	7.21 ± 0.07	7.27 ± 0.15	7.00 ± 0.23	7.16 ± 0.14
9. Free Carbon Dioxide (mg/l)	2.12 ± 0.14	5.05 ± 0.20	1.76 ± 0.08	2.98 ± 1.80	2.10 ± 0.16	4.71 ± 0.36	1.66 ± 0.09	2.82 ± 1.65
10. Dissolved Oxygen (mg/l)	9.71 ± 0.70	7.53 ± 0.24	11.68 ± 0.42	9.64 ± 2.07	10.61 ± 0.93	7.32 ± 0.25	10.82 ± 0.50	9.58 ± 1.96

Table 4A.3. (Contd.)

11. B.O.D. (mg/l)	2.35 ± 0.26	3.00 ± 0.55	1.41 ± 0.11	2.25 ± 0.80	2.35 ± 0.13	2.83 ± 0.59	1.71 ± 0.58	2.30 ± 0.56
12. C.O.D. (mg/l)	9.06 ± 0.98	5.42 ± 2.13	11.57 ± 0.44	8.69 ± 3.09	9.06 ± 0.52	6.57 ± 2.28	10.89 ± 2.22	8.84 ± 2.17
13 Total Alkalinity (mg/l)	66.89 ± 1.62	47.89 ± 0.78	35.22 ± 0.44	50.00 ± 15.94	68.89 ± 0.60	47.22 ± 1.48	37.89 ± 2.93	51.33 ± 15.90
14 Chloride (mg/l)	2.96 ± 0.03	11.39 ± 0.31	1.97 ± 0.04	5.44 ± 5.18	2.88 ± 0.10	11.41 ± 0.57	1.83 ± 0.25	5.38 ± 5.25
15. Total Phosphate (mg/l)	0.02 ± 0.00	0.11 ± 0.01	0.07 ± 0.01	0.07 ± 0.04	0.02 ± 0.00	0.13 ± 0.01	0.07 ± 0.01	0.07 ± 0.05
16. Total Organic Carbon (mg/l)	3.16 ± 0.05	5.23 ± 0.10	4.14 ± 0.03	4.18 ± 1.04	3.36 ± 0.10	5.29 ± 0.13	4.20 ± 0.03	4.28 ± 0.97
17 Hardness (mg/l)	83.33 ± 1.50	105.53 ± 2.21	95.78 ± 0.67	94.88 ± 11.13	85.33 ± 0.87	111.89 ± 2.03	93.33 ± 14.16	96.85 ± 13.62
18. Ions								
(a) Sodium (mg/l)	9.2021 ± 0.01	22.7991 ± 0.02	11.3939 ± 0.02	14.4650 ± 7.30	10.1795 ± 0.02	24.7245 ± 0.01	13.4170 ± 0.02	16.1070 ± 7.64
(b) Potassium (mg/l)	1.2012 ± 0.01	2.6337 ± 0.01	1.5951 ± 0.01	1.8100 ± 0.74	1.2230 ± 0.01	2.6231 ± 0.02	1.6093 ± 0.01	1.8185 ± 0.72
(c) Sulphate (mg/l)	18.8000 ± 0.03	24.6400 ± 0.01	19.5600 ± 0.01	21.0000 ± 3.18	18.9900 ± 0.02	24.6400 ± 0.02	18.3300 ± 0.02	20.6533 ± 3.47
(d) Nitrate (mg/l)	0.0130 ± 0.02	0.0700 ± 0.02	0.0400 ± 0.02	0.0410 ± 0.03	0.0140 ± 0.01	0.0700 ± 0.01	0.0300 ± 0.01	0.0380 ± 0.03

± = Standard Deviation.

Table 4A.4. Seasonal Variation in Heavy Metals of River Ganga at Sampling Station B (1998-2000)

Seasons Parameters	1998 to 1999				1999 to 2000			
	Summer	Monsoon	Winter	Average	Summer	Monsoon	Winter	Average
Lead [Pb] (mg/l)	0.0757 ± 0.0015	0.0780 ± 0.0036	0.0628 ± 0.0011	0.0722 ± 0.0082	0.0776 ± 0.0024	0.0791 ± 0.0014	0.0613 ± 0.0003	0.0727 ± 0.0099
Copper [Cu] (mg/l)	0.0177 ± 0.0005	0.0184 ± 0.0002	0.0158 ± 0.0003	0.0173 ± 0.0013	0.0183 ± 0.0002	0.0187 ± 0.0004	0.0153 ± 0.0002	0.0174 ± 0.0019
Chromium [Cr] (mg/l)	0.0260 ± 0.0036	0.0293 ± 0.0029	0.0306 ± 0.0005	0.0286 ± 0.0024	0.0353 ± 0.0025	0.0357 ± 0.0009	0.0254 ± 0.0004	0.0322 ± 0.0058
Cadmium [Cd] (mg/l)	0.0023 ± 0.0004	0.0027 ± 0.0002	0.0034 ± 0.0002	0.0028 ± 0.0005	0.0033 ± 0.0005	0.0038 ± 0.0002	0.0022 ± 0.0002	0.0031 ± 0.0008
Zinc [Zn] (mg/l)	3.0100 ± 0.0300	3.1150 ± 0.0100	3.1814 ± 0.0523	3.1021 ± 0.0864	3.0641 ± 0.0422	3.2490 ± 0.0157	3.2100 ± 0.0010	3.1744 ± 0.0975

± = Standard Deviation.

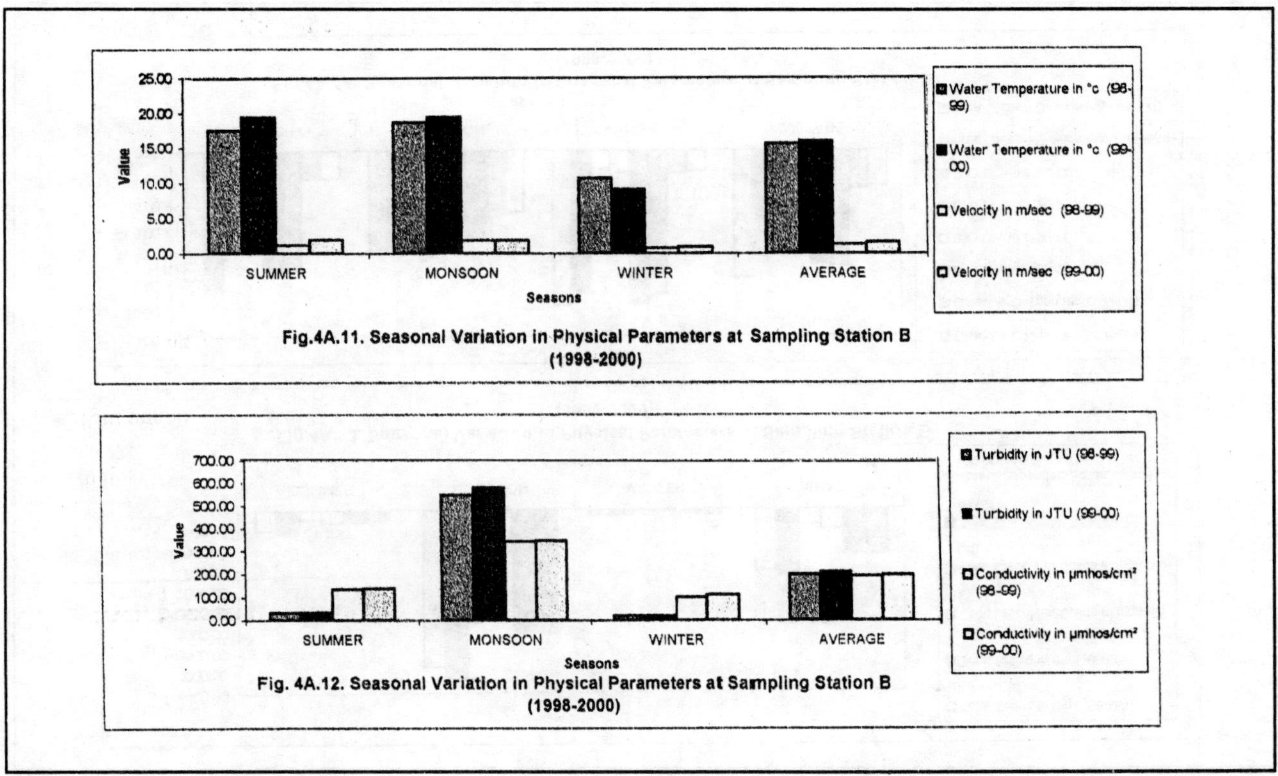

Fig.4A.11. Seasonal Variation in Physical Parameters at Sampling Station B (1998-2000)

Fig. 4A.12. Seasonal Variation in Physical Parameters at Sampling Station B (1998-2000)

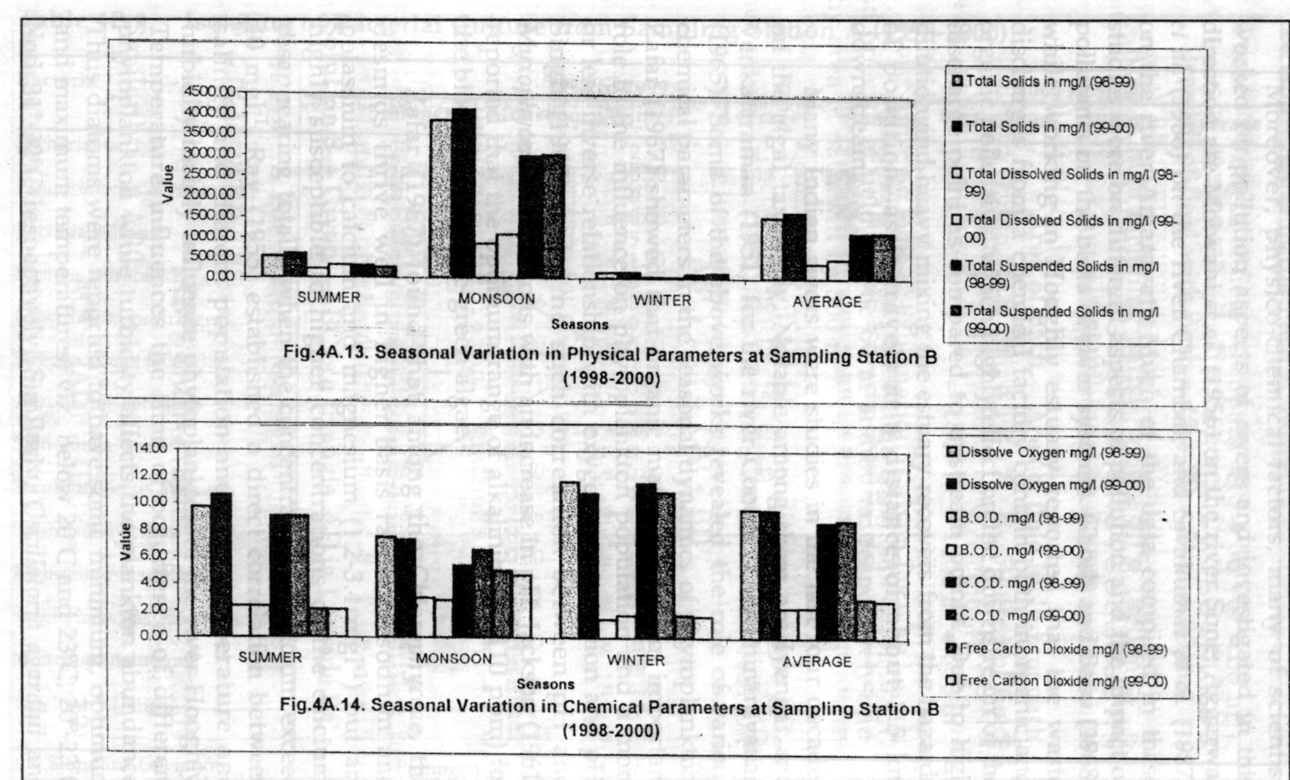

Fig.4A.13. Seasonal Variation in Physical Parameters at Sampling Station B (1998-2000)

Fig.4A.14. Seasonal Variation in Chemical Parameters at Sampling Station B (1998-2000)

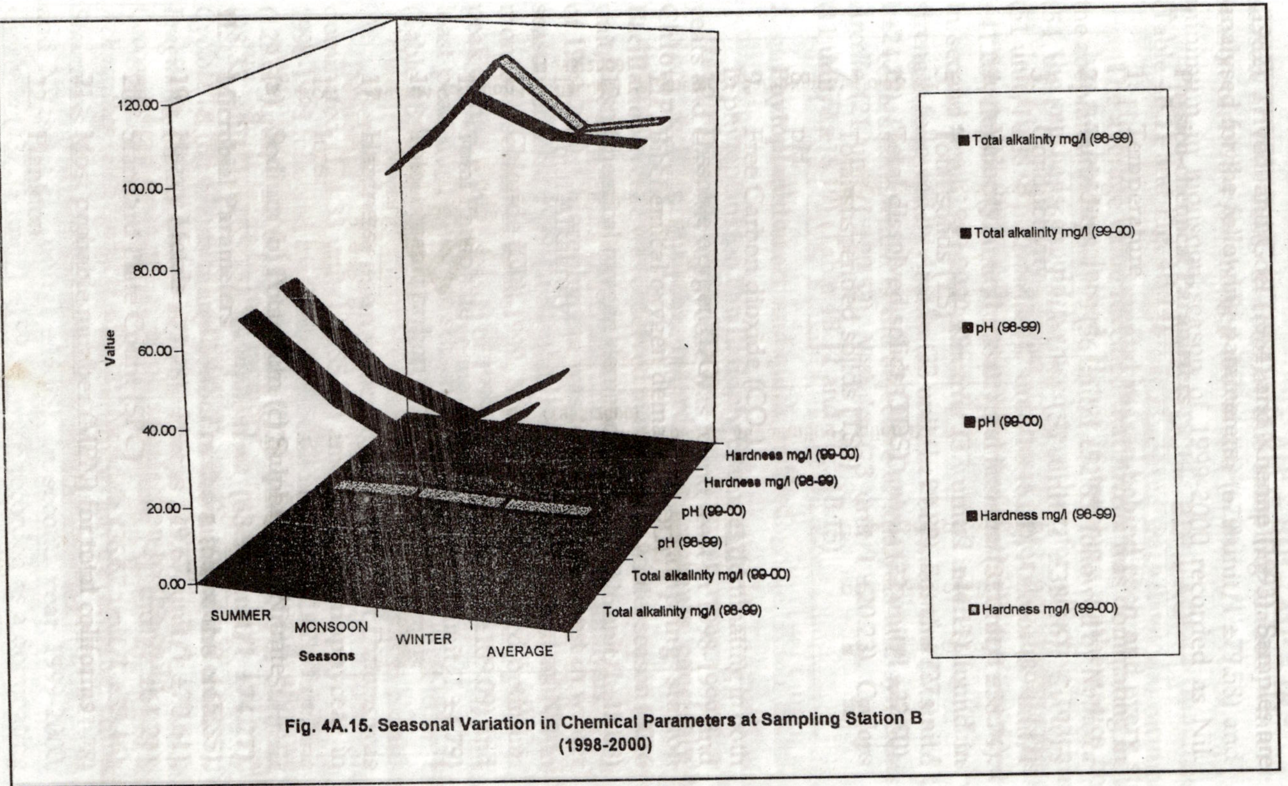

Fig. 4A.15. Seasonal Variation in Chemical Parameters at Sampling Station B (1998-2000)

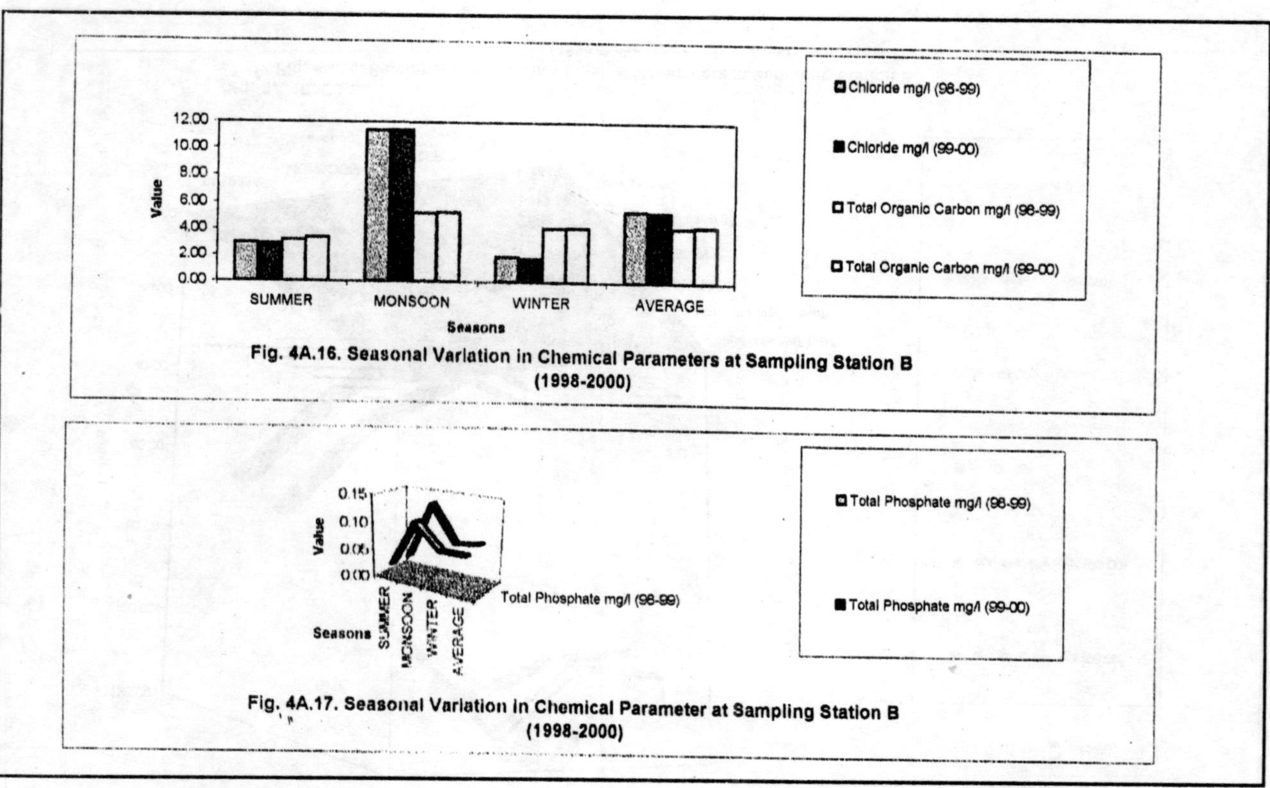

Fig. 4A.16. Seasonal Variation in Chemical Parameters at Sampling Station B (1998-2000)

Fig. 4A.17. Seasonal Variation in Chemical Parameter at Sampling Station B (1998-2000)

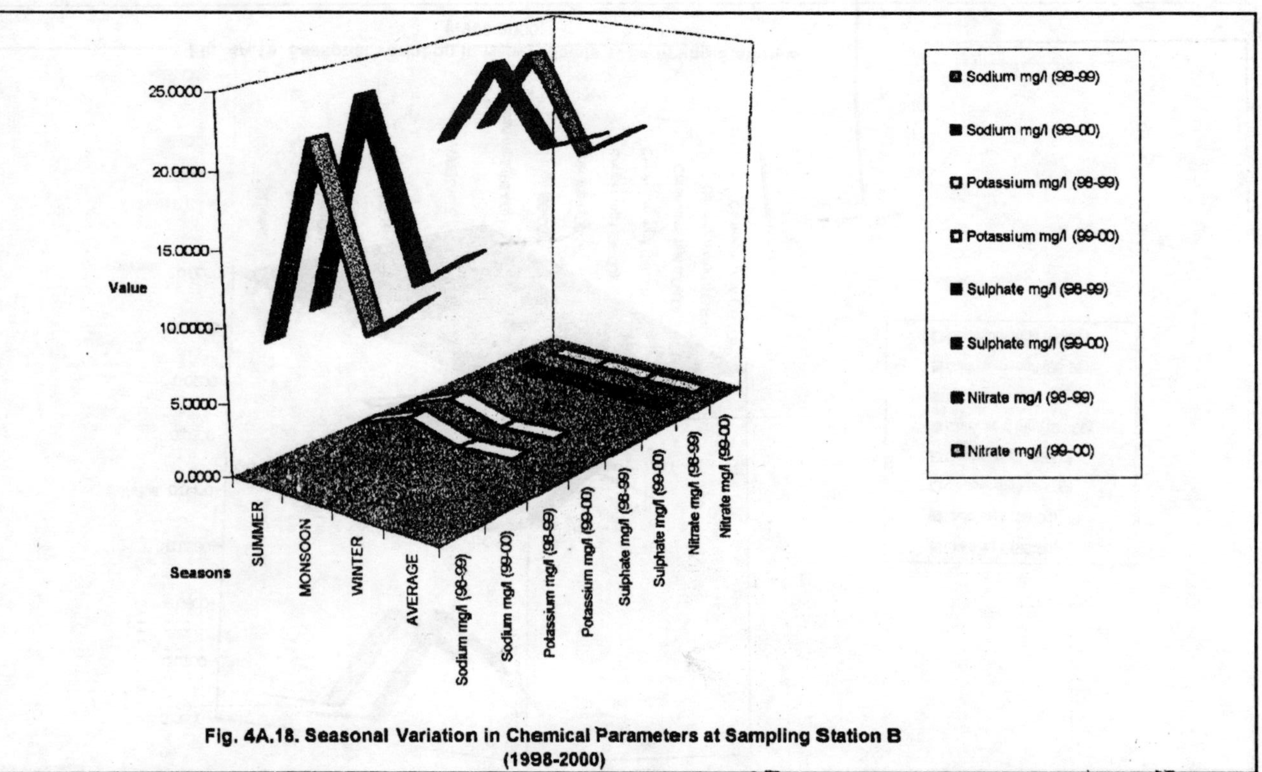

Fig. 4A.18. Seasonal Variation in Chemical Parameters at Sampling Station B (1998-2000)

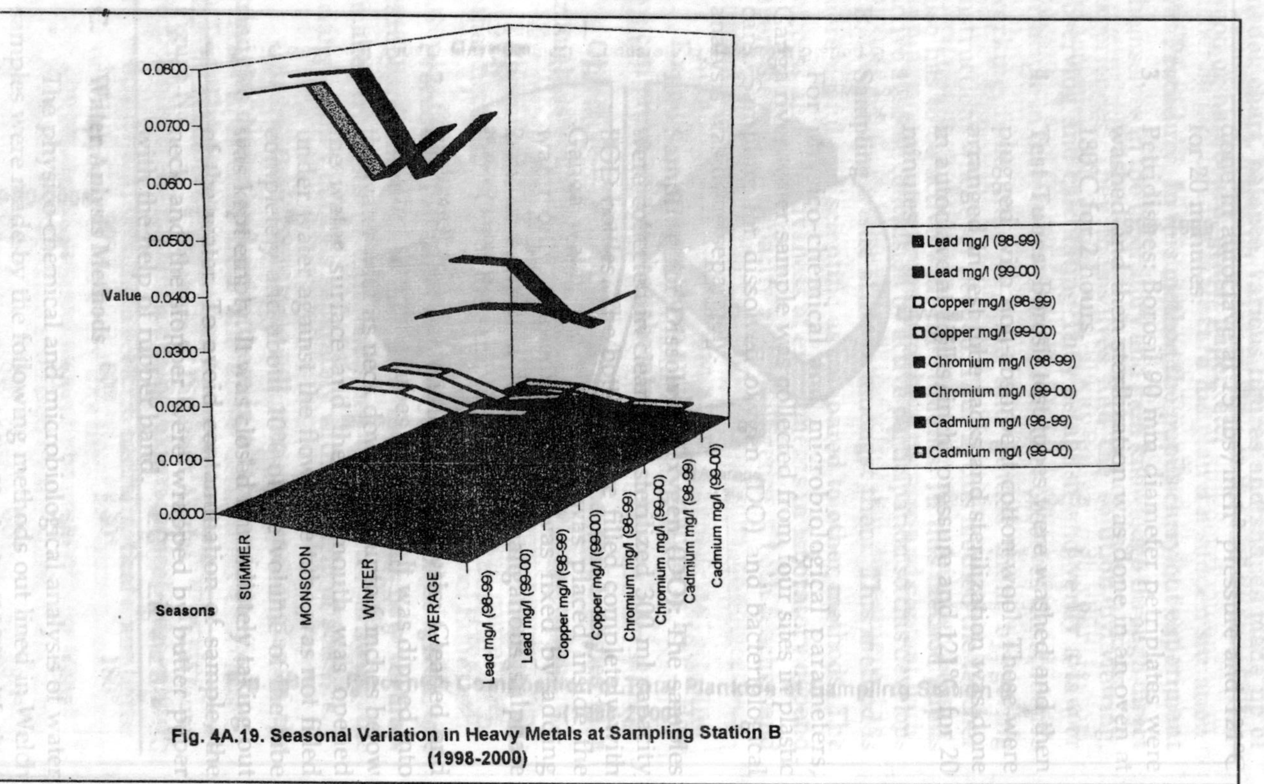

Fig. 4A.19. Seasonal Variation in Heavy Metals at Sampling Station B (1998-2000)

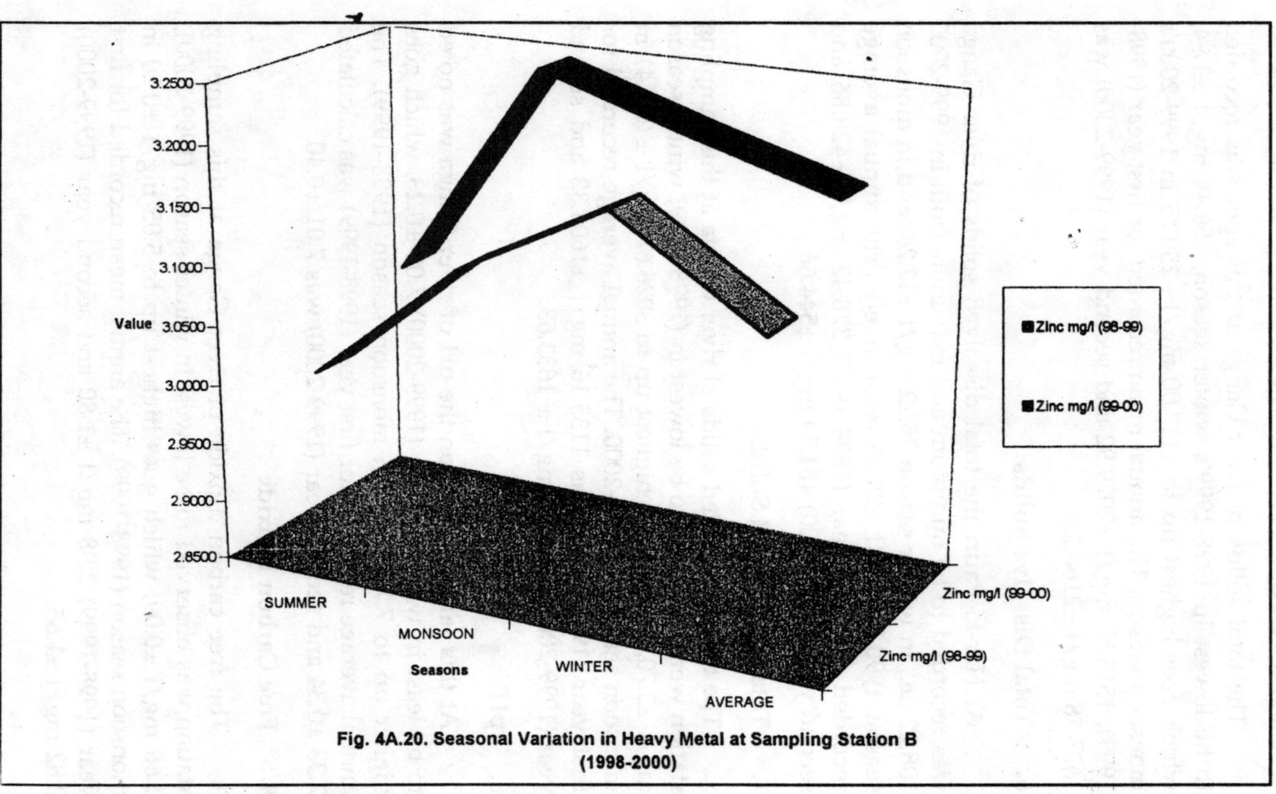

Fig. 4A.20. Seasonal Variation in Heavy Metal at Sampling Station B (1998-2000)

5. Total Solids

The total solids of river Ganga at this spot was recorded to be lowest in 1998-1999's winter season, 156.44 mg/l ±1.94, which goes highest up to 4117.00 mg/l ±250.75 in 1999-2000's monsoon season. The annual mean recorded for first year (1998-1999), 1513.57 mg/l ±2029.92 and second year (1999-2000) was 1625.78 mg/l ±2168.23.

6. Total Dissolve Solids

At Har-ki-Pauri the total dissolved solids of river Ganga was recorded to be minimum and maximum both in 1999-2000 108.22 i.e., in winter season 35.22 mg/l ±12.26, and in monsoon season 1990.33 mg/l ±98.25 respectively. The annual average recorded for first year (1998-1999) 378.42 mg/l ±428.86 and second year (1999-2000) 484.59 mg/l ±544.66.

7. Total Suspended Solids

The total suspended solids of river Ganga at this sampling station were observed to be lowest in (1998-1999) winter season mg/l ±3.70, which goes highest up to 3026.67 mg/l ± 210.41 in monsoon season of 1999-2000. The annual average recorded for first year (1998-1999) was 1135.16 mg/l ±1604.33 and second year (1999-2000) 1142.07 mg/l ± 1633.63.

8. pH

At this sampling station the pH of river Ganga was noted to be least in winter season (1999-2000) 7.00 ±0.23, which goes higher up to 7.71 ±0.14 in monsoon season (1998-1999). The annual average recorded for first year (1998-1999) was calculated 7.33 ±0.34 and second year (1999-2000) was 7.01 ±0.40.

9. Free Carbon Dioxide

The free carbon dioxide of river Ganga at this sampling station was observed to be lowest in winter season (1999-2000), 1.66 mg/l ±0.09, which goes highest up to 5.05 mg/l ±0.20 in monsoon season (1998-1999). The annual mean recorded for first year (1998-1999) 2.98 mg/l ±1.80 and second year (1999-2000) 2.82 mg/l ±1.65.

10. Dissollved Oxygen

The dissolved oxygen of river Ganga at this sampling station was recorded to be minimum in monsoon season (1998-1999) 7.32 mg/l ±0.25, which goes maximum up to 11.68mg/l ±0.42 in winter season (1998-1999). The annual average observed for first year (1998-1999), 9.64 mg/l ±2.07 and second year (1999-2000), was 9.58 mg/l ±1.96.

11. Biological Oxygen Demand (BOD)

In the same year (1998-1999) the biological oxygen demand (BOD) of river Ganga at this spot was noted to be lowest in winter season 1.41 mg/l ±0.11, which goes highest up to 3.00 mg/l ±0.55 in monsoon season. The annual average recorded for first year (1998-1999) 2.25 mg/l ±0.80 and second year (1999-2000) 2.30 mg/l ±0.56.

12. Chemical Oxygen Demand (COD)

The chemical oxygen demand (COD) of river Ganga at this sampling station was observed to be least in monsoon season (1998-1999) 5.42 mg/l ±2.13, which goes highest up to 11.57 mg/l ±0.44 in winter season (1998-1999). The annual mean calculated for first year (1998-1999) 8.69 mg/l ±3.09 and second year (1999-2000) 8.84 mg/l ±2.17.

13. Total Alkalinity

The total alkalinity of water at Har-ki-Pauri was observed to be minimum in winter season (1998-1999), 35.22 mg/l ±0.44, which goes highest up to 68.89 mg/l ±0.60 in summer season (1999-2000). The annual average recorded for first year (1998-1999) was 50.00 mg/l ±15.94 and second year (1999-2000) was 51.33 mg/l ±15.90.

14. Chloride

The chloride in the water of river Ganga at Har-ki-Pauri this spot was observed to be lowest in winter season (1999-2000), 1.83 mg/l ±0.25, which goes highest up to 11.41 mg/l ±0.57 in monsoon season (1999-2000). The annual average recorded for first year (1998-1999) was 5.44 mg/l ±5.18 and second year (1999-2000) was 5.38 mg/l ±5.25.

15. Total Phosphate

At this sampling station the total phosphate of river Ganga was observed to be minimum in summer season of both years, 0.02 mg/l ±0.00, which goes maximum up to 0.13 mg/l ±0.01 in monsoon season (1999-2000). The annual average observed for first year (1998-1999) was 0.07 mg/l ±0.04 and second year (1999-2000) was 0.07 mg/l ±0.05.

16. Total Organic Carbon (TOC)

The total organic carbon (TOC) at this spot of river Ganga was recorded to be lowest in summer season (1998-1999), 3.16 mg/l ±0.05, which goes maximum up to 5.29 mg/l ±0.13 in monsoon season (1999-2000). The annual mean noted for first year (1998-1999) was 4.18 mg/l ±1.04 and second year (1999-2000) was 4.28 mg/l ±0.97.

17. Hardness

The hardness at this sampling station of river Ganga was observed to be least in summer season (1998-1999), 83.33 mg/l ±1.50, which goes maximum up to 111.89 mg/l ±2.03 in monsoon season (1999-2000). The annual average observed for first year (1998-1999) was 94.88 mg/l ±11.13 and second year (1999-2000) was 96.85 mg/l ±13.62.

18. Ions

(a) Sodium

The sodium at this spot of river Ganga was recorded to be minimum in summer season (1998-1999), 9.2021 mg/l ±0.01, which goes maximum up to 24.7245 mg/l ±0.01 in monsoon season (1999-2000). The annual average was found 14.4650 mg/l ±7.30 for first year (1998-1999) and 16.1070 mg/l ±7.64 for second year (1999-2000).

(b) Potassium

At this sampling station the potassium of river Ganga was observed to be lowest in summer season (1998-1999), 1.2012 mg/l ±0.01, which goes highest up to 2.6337 mg/l ±0.01 in monsoon season (1998-1999). The annual mean calculated for first year (1998-1999) was 1.81 mg/l ±0.74 and second year (1999-2000) was 1.8185 mg/l ±0.72.

(c) Sulphate

At this sampling station the sulphate of river Ganga was recorded to be minimum in winter season (1999-2000), 18.33 mg/l ±0.02, which goes maximum up to 24.64 mg/l ±0.01 in monsoon season (1998-1999), and same in monsoon season of 1999-2000. The annual average observed for first year (1998-1999) was 21.00 mg/l ±3.18 and second year (1999-2000) was 20.6533 mg/l ±3.47.

(d) Nitrate

The nitrate in water at Har-ki-Pauri was observed to be lowest in summer season (1998-1999), 0.013 mg/l ±0.02, which goes highest up to 0.07 mg/l ±0.01 in monsoon season of 1998-1999. The annual average recorded for first year (1998-1999) was 0.041 mg/l ±0.03 and second year (1999-2000) was 0.038 mg/l ±0.03.

19. Heavy Metals

(a) Lead (Pb)

The lead at this sampling station of river Ganga was recorded to be minimum in winter season (1999-2000), 0.0613 mg/l ±0.0003, which goes maximum up to 0.0791 mg/l ±0.0014 in monsoon season (1999-2000). The annual average for first year (1998-1999) was 0.0722 mg/l ±0.0082 and second year (1999-2000) was 0.0727 mg/l ±0.0099.

(b) Copper (Cu)

At this spot the copper was recorded to be lowest in winter season (1999-2000), 0.0153 mg/l ±0.0002, which goes highest up to 0.0187 mg/l ±0.0004 in monsoon season (1999-2000). The annual average observed for first year (1998-1999) was 0.0173 mg/l ±0.0013 and second year (1999-2000) was 0.0174 mg/l ±0.0019.

(c) Chromium (Cr)

At this sampling station the chromium of river Ganga was recorded to be least in summer season (1998-1999), 0.0260 mg/l ±0.0036, which goes optimum up to 0.0357 mg/l ±0.0009 in monsoon season (1999-2000). The annual average calculated for first year (1998-1999) was 0.0286 mg/l ±0.0024 and second year (1999-2000) was 0.0322 mg/l ±0.0058.

(d) Cadmium (Cd)

The cadmium at this sampling station of river Ganga was observed to be lowest in winter season (1999-2000), 0.0022 mg/l ±0.0002, which goes highest up to 0.0038 mg/l ±0.0002 in monsoon season (1999-2000). The annual average observed for first year (1998-1999) was 0.0028 mg/l ±0.0005 and second year (1999-2000) was 0.0031 mg/l ±0.0008.

(e) Zinc (Zn)

At this spot the zinc was recorded to be minimum in summer season (1998-1999), 3.01 mg/l ±0.03, which goes maximum up to 3.2490 mg/l ±0.0157 in monsoon season (1999-2000). The annual mean calculated for first year (1998-1999) was 3.1021 mg/l ±0.0864 and second year (1999-2000) was 3.1744 mg/l ±0.975.

SAMPLING STATION C (Muneshwar Ghat)

Physico-chemical Parameters: The physical parameters of the sampling station C are presents in Table 4A.5 and Figs. 4A.21 to 4A.23 and the chemical Parameters are in Tables 4A.5-4A.6 and Figs. 4A.24-4A.30.

1. Water Temperature

The water temperature of river Ganga at this sampling station during total study period was observed to be lowest in winter season (1999-2000), 8.83°C ±0.86, which goes highest up to 20.11°C ±0.44 in monsoon season (1999-2000). The annual average recorded for first year (1998-1999) 15.41°C ±4.35 and for second year (1999-2000) it was calculated 16.31°C ±6.48.

2. Turbidity

The turbidity of water at this spot observed to be minimum in 1999-2000 winter season 19.11 JTU ±2.20, which goes maximum up to 597 JTU ±26.47 in 1999-2000 monsoon season. The annual mean recorded for first year (1998-1999) was 198.05 JTU ±296.80 and second year (1999-2000) was 217.13 JTU ±329.08.

3. Conductivity

The conductivity of water at Muneshwar Ghat was observed to be least in winter season of 1998-1999, 99.33 μmhos/cm^2 ±2.18, which goes highest up to 367.11 μmhos/cm^2 ±23.56 in monsoon season of 1999-2000. The annual average observed for first year

(1998-1999) was 193.33 μmhos/cm^2 ±130.59 and second year (1999-2000) was 208.41 μmhos/cm^2 ±137.84.

4. Velocity

The velocity of this sampling station at river Ganga was noted to be minimum in winter season of 1998-1999, 0.73 m/sec ±0.05, which goes maximum up to 1.94 m/sec ±0.04 in monsoon season of 1999-2000. The annual average calculated for first year (1998-1999) was 1.23 m/sec ±0.57 and second year (1999-2000) was 1.50 m/sec ±0.53.

5. Total Solids

The total solids of river Ganga at this spot was recorded to be lowest in 1998-1999's winter season 157 mg/l ±2.87, which goes highest up to 3927.44 mg/l ±275.94 in 1999-2000's monsoon season. The annual mean recorded for first year (1998-1999), 1506.50 mg/l ±2017.51 and second year (1999-2000) was 1566.81 mg/l ±2055.99.

6. Total Dissolve Solids

At Muneshwar Ghat the total dissolved solids of river Ganga was recorded to be minimum and maximum both in 1999-2000 i.e., in winter season 31.00 mg/l ±4.86, and in monsoon season 973.33 mg/l ±126.33 respectively. The annual average recorded for first year (1998-1999) 376.84 mg/l ±426.07 and second year (1999-2000) 431.70 mg/l ±486.69.

7. Total Suspended Solids

The total suspended solids of river Ganga at this sampling station were observed to be lowest and highest in (1998-1999). Lowest was in winter season 108.44 mg/l ±4.33, which goes highest up to 2967.44 mg/l ±222.01 in monsoon season of 1999-2000. The annual average recorded for first year (1998-1999) was 1129.65 mg/l ±1594.86 and second year (1999-2000) 1133.52 mg/l ±1579.03.

8. pH

At this sampling station the pH of river Ganga was noted to be least and optimum both in 1998-1999. The least was in winter season 7.06 ±0.10, which goes optimum up to 7.53 ±0.19 in monsoon season. The annual average recorded for first year (1998-1999) was calculated 7.28 ±0.24 and second year (1999-2000) was 7.24 ±0.01.

Table 4A.5. Seasonal Variation in Physico-chemical Parameter of River Ganga at Sampling Station C (1998-2000)

Physico-chemical Parameters	1998 to 1999				1999 to 2000			
	Summer	Monsoon	Winter	Average	Summer	Monsoon	Winter	Average
1. Water Temperature (°C)	17.22 ± 0.67	18.56 ± 0.46	10.44 ± 0.63	15.41 ± 4.35	20.00 ± 0.56	20.11 ± 0.44	8.83 ± 0.86	16.31 ± 6.48
2. Turbidity (JTU)	34.05 ± 5.47	540.67 ± 15.55	19.44 ± 1.59	198.05 ± 296.80	35.28 ± 3.17	597.00 ± 26.47	19.11 ± 2.20	217.13 ± 329.08
3. Conductivity (μmhos/cm^2)	138.22 ± 1.48	342.44 ± 8.50	99.33 ± 2.18	193.33 ± 130.59	139.56 ± 2.40	367.11 ± 23.56	118.56 ± 5.32	208.41 ± 137.84
4. Velocity (m/sec)	1.12 ± 0.01	1.86 ± 0.01	0.73 ± 0.05	1.23 ± 0.57	1.64 ± 0.05	1.94 ± 0.04	0.91 ± 0.04	1.50 ± 0.53
5. Total Solids (mg/l)	536.71 ± 28.17	3825.78 ± 226.43	157.00 ± 2.87	1506.50 ± 2017.51	604.78 ± 63.29	3927.44 ± 275.94	168.22 ± 13.56	1566.81 ± 2055.99
6. Total Dissolved Solids (mg/l)	223.64 ± 8.49	858.33 ± 14.98	48.56 ± 4.42	376.84 ± 426.07	290.67 ± 50.26	973.33 ± 126.33	31.11 ± 4.86	431.70 ± 486.69
7. Total Suspended Solids (mg/l)	313.07 ± 33.06	2967.44 ± 222.01	108.44 ± 4.33	1129.65 ± 1594.86	309.33 ± 27.68	2954.11 ± 224.09	137.11 ± 14.37	1133.52 ± 1579.03
8. pH	7.27 ± 0.17	7.53 ± 0.19	7.06 ± 0.10	7.28 ± 0.24	7.25 ± 0.15	7.24 ± 0.14	7.24 ± 0.18	7.24 ± 0.01
9. Free Carbon Dioxide (mg/l)	2.21 ± 0.29	4.95 ± 0.24	1.75 ± 0.07	2.97 ± 1.73	2.21 ± 0.20	4.76 ± 0.38	1.68 ± 0.15	2.88 ± 1.65
10. Dissolved Oxygen (mg/l)	9.52 ± 0.89	7.28 ± 0.26	11.74 ± 0.43	9.51 ± 2.23	10.69 ± 0.63	7.55 ± 0.31	10.94 ± 0.51	9.73 ± 1.89

Table 4A.5. (Contd.)

11. B.O.D. (mg/l)	2.36 ± 0.27	3.03 ± 0.53	1.38 ± 0.11	2.26 ± 0.83	2.34 ± 0.13	2.76 ± 0.29	1.94 ± 0.56	2.35 ± 0.41
12. C.O.D. (mg/l)	9.09 ± 1.02	5.33 ± 2.03	11.65 ± 0.41	8.69 ± 3.18	9.03 ± 0.51	7.49 ± 1.13	10.63 ± 2.14	9.05 ± 1.57
13 Total Alkalinity (mg/l)	66.67 ± 2.50	48.22 ± 0.67	35.67 ± 1.00	50.19 ± 15.59	67.22 ± 0.67	49.11 ± 1.36	38.78 ± 2.17	51.70 ± 14.40
14 Chloride (mg/l)	2.96 ± 0.03	11.43 ± 0.13	1.96 ± 0.04	5.45 ± 5.20	2.88 ± 0.10	11.31 ± 0.21	1.82 ± 0.27	5.34 ± 5.20
15. Total Phosphate (mg/l)	0.02 ± 0.00	0.11 ± 0.01	0.07 ± 0.01	0.07 ± 0.04	0.03 ± 0.00	0.13 ± 0.01	0.07 ± 0.01	0.07 ± 0.05
16. Total Organic Carbon (mg/l)	3.37 ± 0.27	5.26 ± 0.13	4.15 ± 0.02	4.26 ± 0.95	3.63 ± 0.10	5.56 ± 0.02	4.30 ± 0.20	4.50 ± 0.98
17 Hardness (mg/l)	83.78 ± 1.30	106.43 ± 2.01	96.22 ± 1.20	95.48 ± 11.35	90.22 ± 1.56	112.48 ± 1.94	93.89 ± 8.59	98.86 ± 11.93
18. Ions								
(*a*) Sodium (mg/l)	8.7198 ± 0.01	22.7991 ± 0.01	11.4110 ± 0.03	14.3100 ± 7.47	10.3927 ± 0.01	26.4991 ± 0.02	13.3753 ± 0.02	16.7557 ± 8.57
(*b*) Potassium (mg/l)	1.1936 ± 0.02	2.6032 ± 0.02	1.6362 ± 0.02	1.8110 ± 0.72	1.1969 ± 0.01	2.6882 ± 0.02	1.5736 ± 0.01	1.8196 ± 0.78
(*c*) Sulphate (mg/l)	18.3000 ± 0.01	23.5600 ± 0.01	19.5000 ± 0.02	20.4533 ± 2.76	17.9800 ± 0.03	23.5600 ± 0.02	17.9800 ± 0.02	19.8400 ± 3.22
(*d*) Nitrate (mg/l)	0.0150 ± 0.01	0.0500 ± 0.01	0.0300 ± 0.01	0.0317 ± 0.02	0.0130 ± 0.01	0.0500 ± 0.01	0.0150 ± 0.01	0.0260 ± 0.02

± = Standard Deviation.

Table 4A.6. Seasonal Variation in Heavy Metals of River Ganga at Sampling Station C (1998-2000)

Seasons Parameters	1998 to 1999				1999 to 2000			
	Summer	Monsoon	Winter	Average	Summer	Monsoon	Winter	Average
Lead [Pb] (mg/lt.)	0.0660 ± 0.0017	0.0757 ± 0.0042	0.0636 ± 0.0039	0.0684 ± 0.0064	0.0770 ± 0.0020	0.0794 ± 0.0009	0.0633 ± 0.0003	0.0732 ± 0.0087
Copper [Cu] (mg/lt.)	0.0684 ± 0.0897	0.0736 ± 0.0956	0.0171 ± 0.0007	0.0530 ± 0.0312	0.0721 ± 0.0926	0.0186 ± 0.0004	0.0156 ± 0.0003	0.0354 ± 0.0318
Chromium [Cr] (mg/lt.)	0.0243 ± 0.0035	0.0290 ± 0.0010	0.0328 ± 0.0025	0.0287 ± 0.0042	0.0280 ± 0.0017	0.0345 ± 0.0016	0.0265 ± 0.0005	0.0297 ± 0.0043
Cadmium [Cd] (mg/lt.)	0.0024 ± 0.0003	0.0030 ± 0.0001	0.0033 ± 0.0002	0.0029 ± 0.0005	0.0034 ± 0.0004	0.0038 ± 0.0002	0.0027 ± 0.0002	0.0033 ± 0.0006
Zinc [Zn] (mg/lt.)	3.1333 ± 0.0208	3.1462 ± 0.0185	3.2240 ± 0.0271	3.1679 ± 0.0491	3.1817 ± 0.0118	3.2470 ± 0.0152	3.2052 ± 0.0037	3.2113 ± 0.0331

± = Standard Deviation.

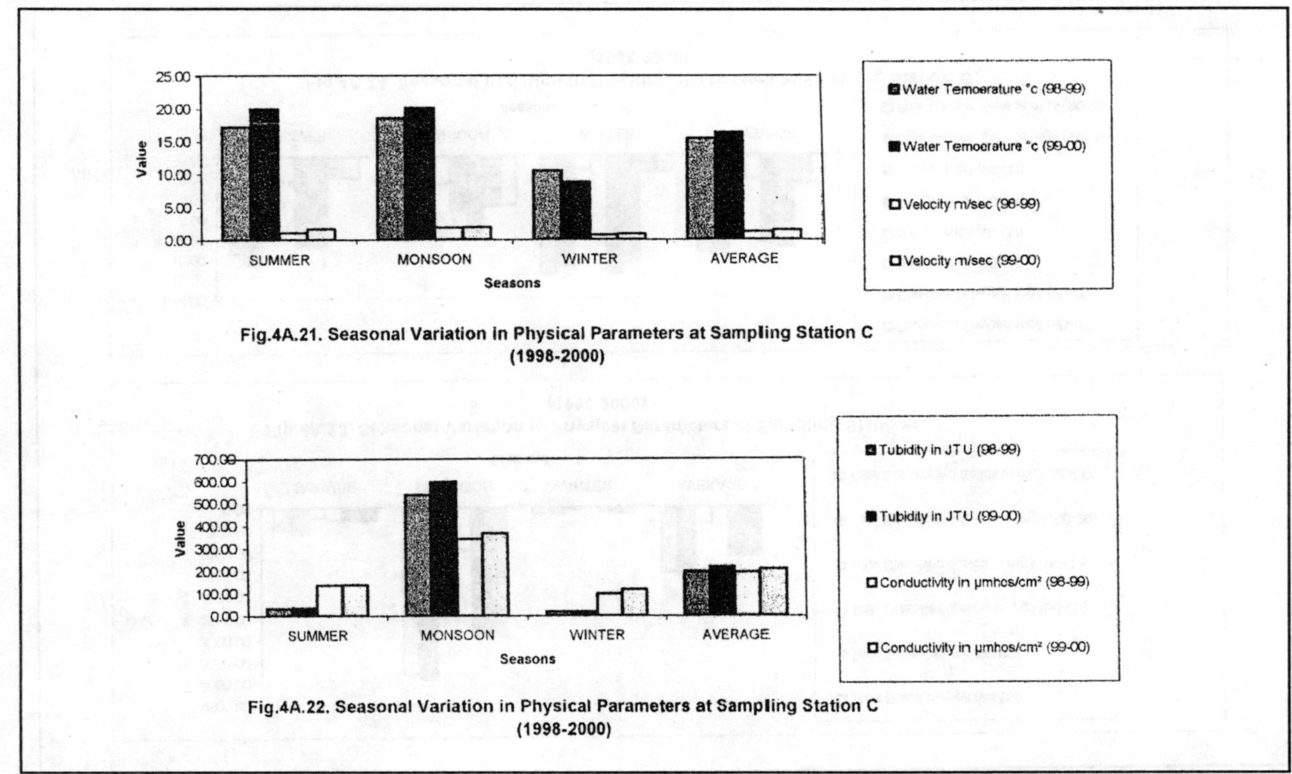

Fig.4A.21. Seasonal Variation in Physical Parameters at Sampling Station C (1998-2000)

Fig.4A.22. Seasonal Variation in Physical Parameters at Sampling Station C (1998-2000)

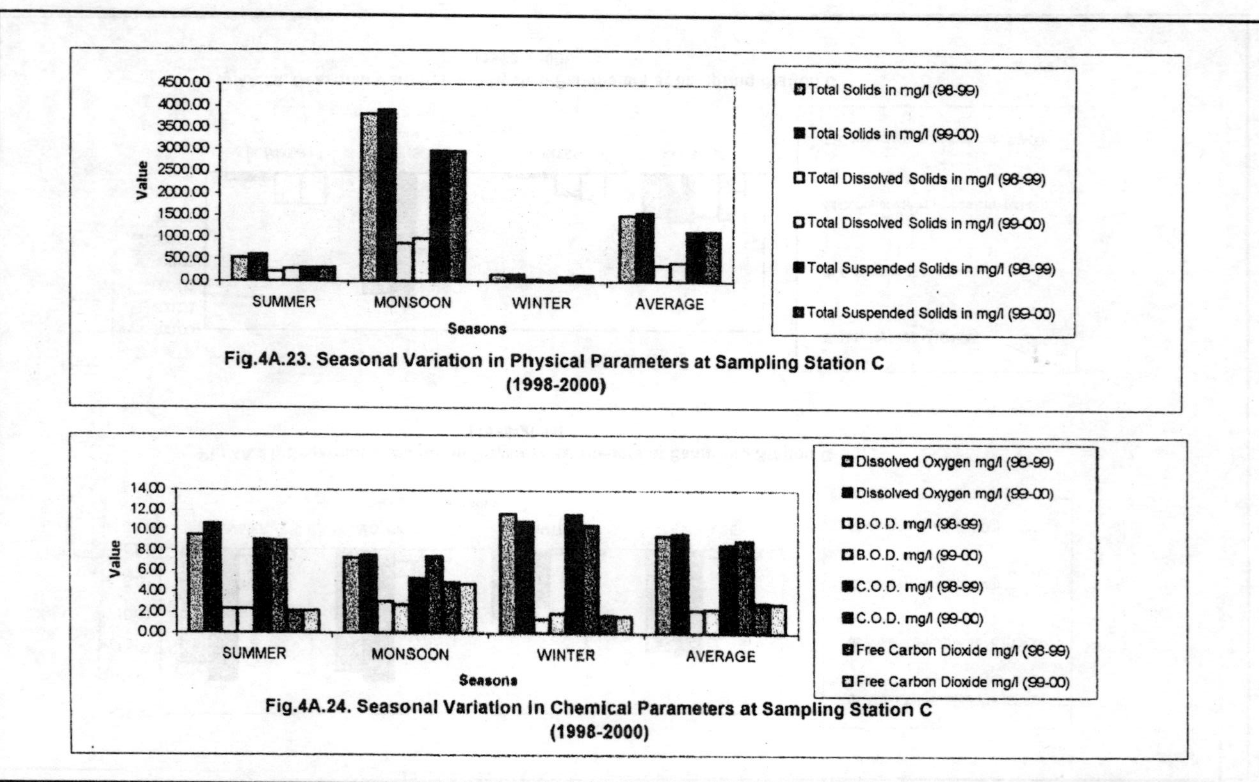

Fig.4A.23. Seasonal Variation in Physical Parameters at Sampling Station C (1998-2000)

Fig.4A.24. Seasonal Variation in Chemical Parameters at Sampling Station C (1998-2000)

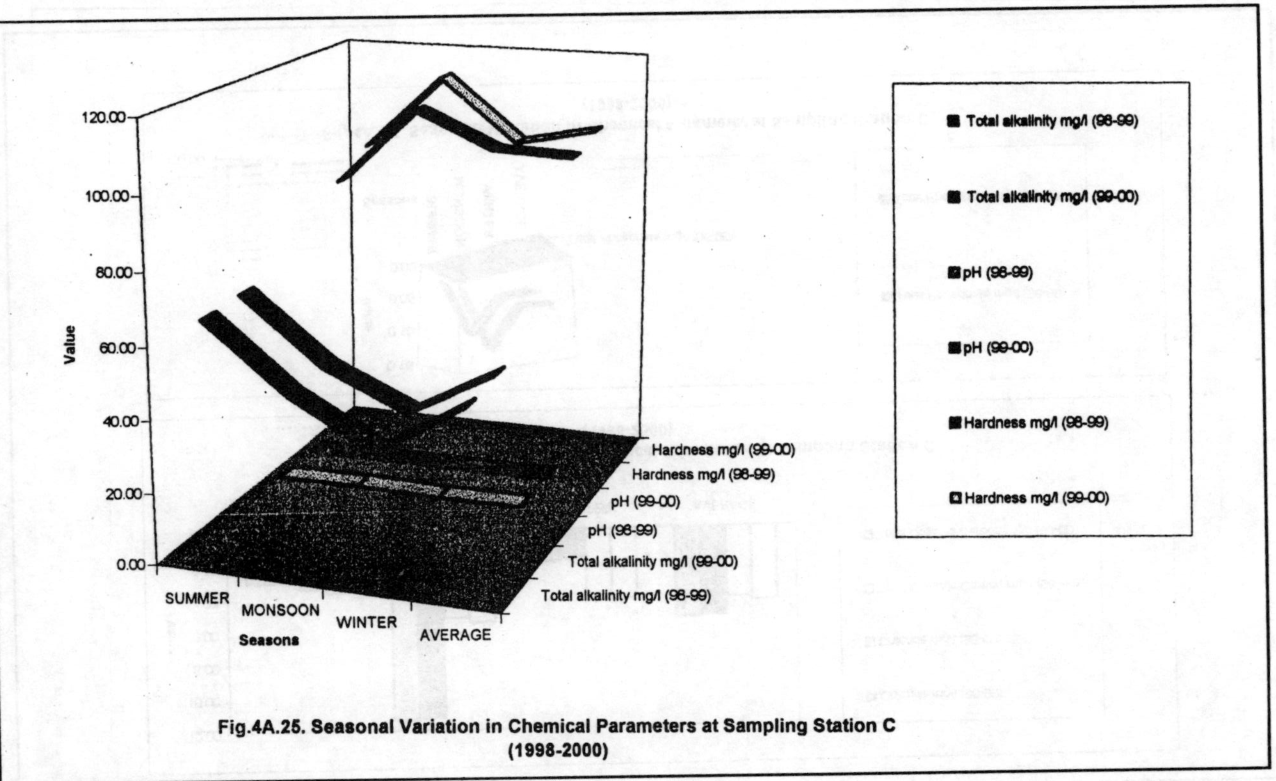

Fig.4A.25. Seasonal Variation in Chemical Parameters at Sampling Station C (1998-2000)

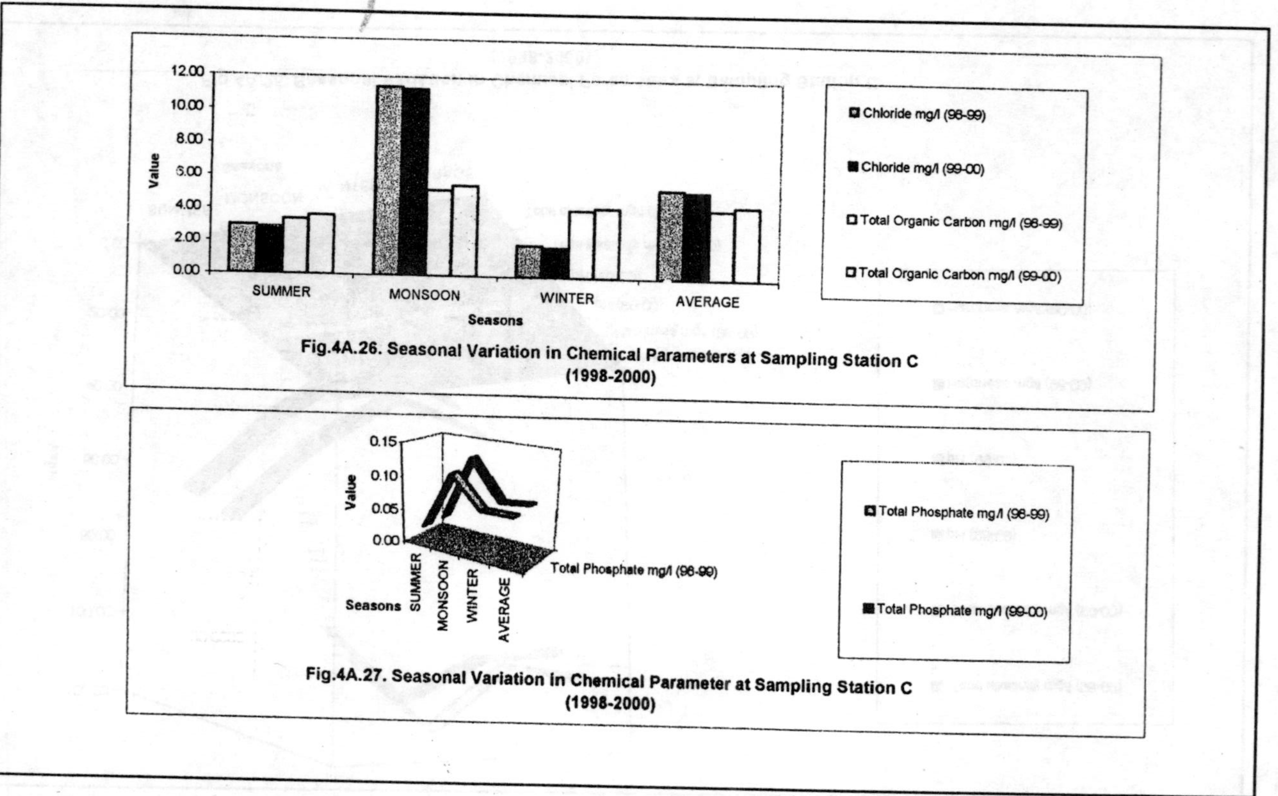

Fig.4A.26. Seasonal Variation in Chemical Parameters at Sampling Station C (1998-2000)

Fig.4A.27. Seasonal Variation in Chemical Parameter at Sampling Station C (1998-2000)

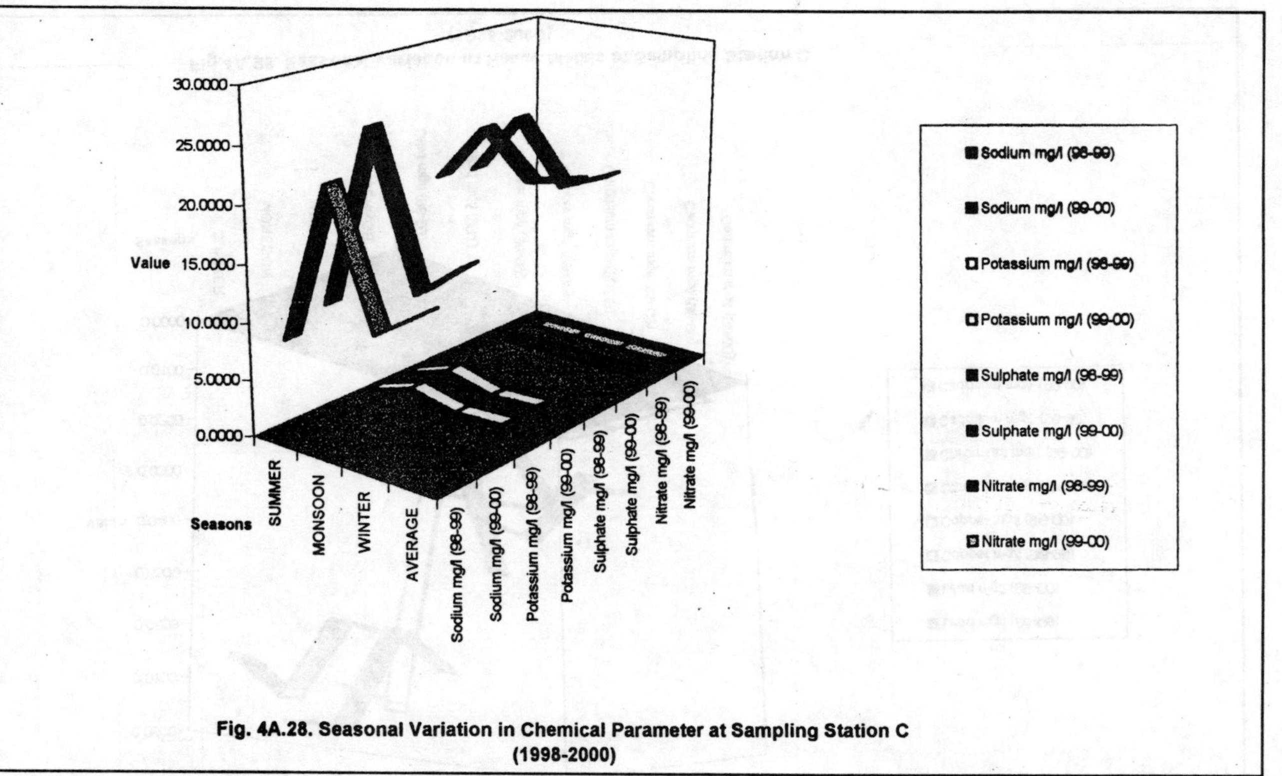

Fig. 4A.28. Seasonal Variation in Chemical Parameter at Sampling Station C (1998-2000)

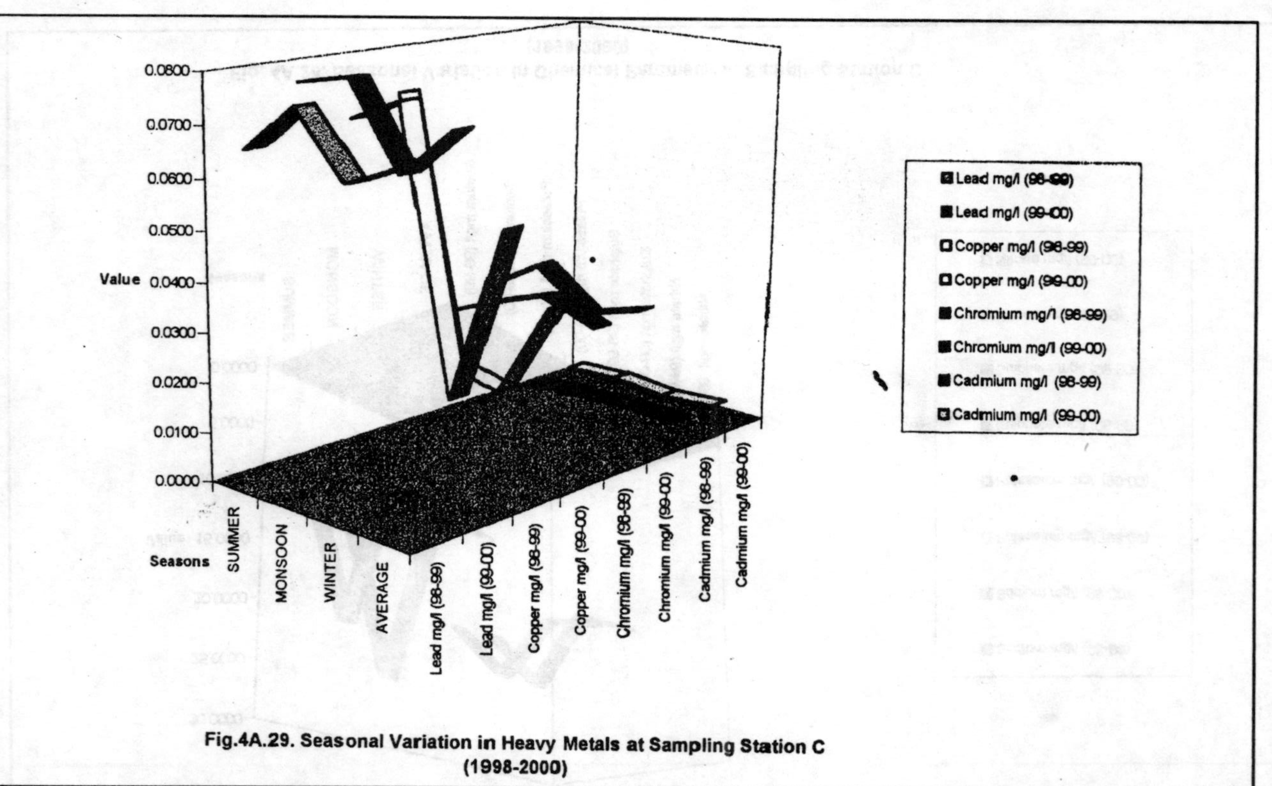

Fig.4A.29. Seasonal Variation in Heavy Metals at Sampling Station C (1998-2000)

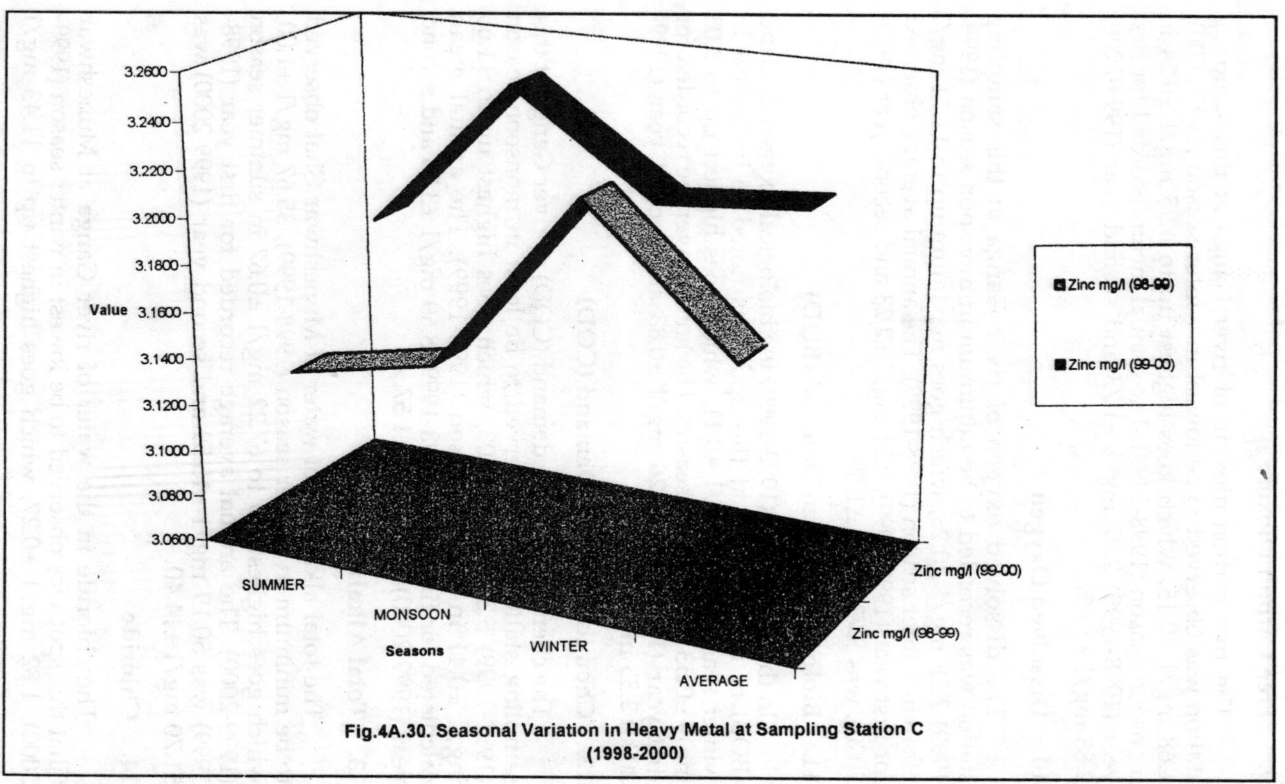

Fig.4A.30. Seasonal Variation in Heavy Metal at Sampling Station C (1998-2000)

9. Free Carbon Dioxide

The free carbon dioxide of river Ganga at this sampling station was observed to be lowest in winter season (1999-2000), 1.68 mg/l ±0.15, which goes highest up to 4.95 mg/l ±0.24 in monsoon season (1998-1999). The annual mean recorded for first year (1998-1999) 2.97 mg/l ±1.73 and second year (1999-2000) 2.88 mg/l ±1.65.

10. Dissolved Oxygen

The dissolved oxygen of river Ganga at this sampling station was recorded to be minimum in monsoon season (1998-1999) 7.28 mg/l ±0.26, which goes maximum up to 11.74 mg/l ±0.43 in winter season (1998-1999). The annual average observed for first year (1998-1999), 9.51 mg/l ±2.23 and second year (1999-2000), was 9.73 mg/l ±1.89.

11. Biological Oxygen Demand (BOD)

In the same year (1998-1999) the biological oxygen demand (BOB) of river Ganga at this spot was noted to be lowest in winter season 1.38 mg/l ±0.11, which goes highest up to 3.03 mg/l ±0.53 in monsoon season. The annual average recorded for first year (1998-1999) 2.26 mg/l ±0.83 and second year (1999-2000) 2.35 mg/l ±0.41.

12. Chemical Oxygen Demand (COD)

The chemical oxygen demand (COD) of river Ganga at this sampling station was observed to be least in monsoon season (1998-1999) 5.33 mg/l ±2.03, which goes highest up to 11.65 mg/l ±0.41 in winter season (1998-1999). The annual mean calculated for first year (1998-1999) 8.69 mg/l ±3.18 and second year (1999-2000) 9.05 mg/l ±1.57.

13. Total Alkalinity

The total alkalinity of water at Muneshwar Ghat observed to be minimum in winter season (1998-1999), 35.67 mg/l ±1.00, which goes highest up to 67.22 mg/l ±0.67 in summer season (1999-2000). The annual average recorded for first year (1998-1999) was 50.19 mg/l ±15.59 and second year (1999-2000) was 51.70 mg/l ±14.40.

14. Chloride

The chloride in the water of river Ganga at Muneshwar Ghat this spot was observed to be lowest in winter season (1999-2000), 1.82 mg/l ±0.27, which goes highest up to 11.43 mg/l

±0.13 in monsoon season (1998-1999). The annual average recorded for first year (1998-1999) was 5.45 mg/l ±5.20 and second year (1999-2000) was 5.34 mg/l ±5.20.

15. Total Phosphate

At this sampling station the total phosphate of river Ganga was observed to be minimum in summer season (1998-1999), 0.02 mg/l ±0.00, which goes maximum up to 0.13 mg/l ±0.01 in monsoon season (1999-2000). The annual average observed for first year (1998-1999) was 0.07 mg/l ±0.04 and second year (1999-2000) was 0.07 mg/l ±0.05.

16. Total Organic Carbon (TOC)

The total organic carbon (TOC) at this spot of river Ganga was recorded to be lowest in summer season (1998-1999), 3.37 mg/l ±0.27, which goes maximum up to 5.56 mg/l ±0.02 in monsoon season (1999-2000). The annual mean noted for first year (1998-1999) was 4.26 mg/l ±0.95 and second year (1999-2000 was 4.50 mg/l ±0.98.

17. Hardness

The hardness at this sampling station of river Ganga was observed to be least in sumer season (1998-1999), 83.78 mg/l ±1.30, which goes maximum up to 112.48 mg/l ±1.94 in monsoon season (1999-2000). The annual average observed for first year (1998-1999) was 95.48 mg/l ±11.35 and second year (1999-2000) was 98.86 mg/l ±11.93.

18. Ions

(a) Sodium

The sodium at this spot of river Ganga was recorded to be minimum in summer season (1998-1999), 8.7198 mg/l ±0.01, which goes maximum up to 26.4991 mg/l ±0.02 in monsoon season (1999-2000). The annual average was found 14.3100 mg/l ±7.47 for first year (1998-1999) and 16.7557 mg/l ±8.57 for second year (1999-2000).

(b) Potassium

At this sampling station the potassium of river Ganga was observed to be lowest in summer season (1998-1999), 1.1936 mg/l ±0.02, which goes highest up to 2.6032 mg/l ±0.02 in monsoon season (1998-1999). The annual mean calculated for first year (1998-1999) was 1.8110 mg/l ±0.72 and second year (1999-2000) was 1.8196 mg/l ±0.78.

(c) *Sulphate*

At this sampling station the sulphate of river Ganga was recorded to be minimum in summer season (1999-2000), 17.9800 mg/l ± 0.03, which goes maximum up to 23.5600 mg/l ±0.01 in monsoon season (1998-1999), and same in monsoon season of 1999-2000. The annual average observed for first year (1998-1999) was 20.4533 mg/l ±2.76 and second year (1999-2000) was 19.8400 mg/l ±3.22.

(d) *Nitrate*

The nitrate in water at Muneshwar Ghat was observed to be lowest in summer season (1999-2000), 0.013 mg/l ±0.01, which goes highest up to 0.05 mg/l ±0.01 in monsoon season of 1998-1999 and 1999-2000. The annual average recorded for first year (1998-1999) was 0.0317 mg/l ±0.02 and second year (1999-2000) was 0.0260 mg/l ±0.02.

19. Heavy Metals

(a) *Lead (Pb)*

The lead at this sampling station of river Ganga was recorded to be minimum in winter season (1999-2000), 0.0633 mg/l ±0.0003, which goes maximum up to 0.0794 mg/l ±0.0009 in monsoon season (1999-2000). The annual average for first year (1998-1999) was 0.0684 mg/l ±0.0064 and second year (1999-2000) was 0.0732 mg/l ±0.0087.

(b) *Copper (Cu)*

At this spot the copper was recorded to be lowest in winter season (1999-2000), 0.0156 mg/l ±0.0003, which goes highest up to 0.0736 mg/l ±0.0956 in monsoon season (1998-1999). The annual average observed for first year (1998-1999) was 0.0530 mg/l ±0.0312 and second year (1999-2000) was 0.0354 mg/l ±0.0318.

(c) *Chromium (Cr)*

At this sampling station the chromium of river Ganga was recorded to be least in summer season (1998-1999), 0.0243 mg/l ±0.0035, which goes optimum up to 0.0345 mg/l ±0.0016 in monsoon season (1999-2000). The annual average calculated for first year (1998-1999) was 0.0287 mg/l ±0.0042 and second year (1999-2000) was 0.0297 mg/l ±0.0043.

(d) Cadmium (Cd)

The cadmium at this sampling station of river Ganga was observed to be lowest in summer season (1998-1999), 0.0024 mg/l ±0.0003, which goes highest up to 0.0038 mg/l ±0.0002 in monsoon season (1999-2000). The annual average observed for first year (1998-1999) was 0.0029 mg/l ±0.0005 and second year (1999-2000) was 0.0033 mg/l ±0.0006.

(e) Zinc (Zn)

At this spot the zinc was recorded to be minimum in summer season (1998-1999), 3.1333 mg/l ±0.0208, which goes maximum up to 3.2240 mg/l ±0.0271 in winter season (1998-1999). The annual mean calculated for first year (1998-1999) was 3.1679 mg/l ±0.0491 and second year (1999-2000) was 3.2113 mg/l ±0.0331.

SAMPLING STATION D (Pul-Jatwara)

Physico-chemical Parameters: The physical parameters of the sampling station D are presents in Table 4A.7 and Figs. 4A.31 to 4A.33 and the chemical parameters are in Tables 4A.7–4A.8 and Figs. 4A.34–4A.40.

1. Water Temperature

The water temperature of river Ganga at this sampling station during total study period was observed to be lowest in winter season (1999-2000), 9.33°C ±0.75, which goes highest up to 20.37°C ±0.22 in monsoon season (1990-2000). The annual average recorded for first year (1998-1999) 15.61°C ±4.53 and for second year (1999-2000) it was calculated 16.57°C ±6.27.

2. Turbidity

The turbidity of water at this spot observed to be minimum in 1998-1999 winter season 19.33 JTU ±1.50, which goes maximum up to 564.33 JTU ±17.99 in 1999-2000 monsoon season. The annual mean recorded for first year (1998-1999) was 198.35 JTU ±298.98 and second year (1999-2000) was 206.85 JTU ±309.65.

Table 4A.7. Seasonal Variation in Physico-chemical Parameter of River Ganga at Sampling Station D (1998-2000)

Physico-chemical Parameters	1998 to 1999				1999 to 2000			
	Summer	Monsoon	Winter	Average	Summer	Monsoon	Winter	Average
1. Water Temperature (°C)	18.00 ± 0.71	18.44 ± 0.81	10.39 ± 0.60	15.61 ± 4.53	20.00 ± 0.35	20.37 ± 0.22	9.33 ± 0.75	16.57 ± 6.27
2. Turbidity (JTU)	32.22 ± 3.84	543.50 ± 16.17	19.33 ± 1.50	198.35 ± 298.98	34.44 ± 2.98	564.33 ± 17.99	21.78 ± 2.11	206.85 ± 309.65
3. Conductivity (μmhos/cm^2)	138.00 ± 1.00	336.78 ± 6.87	99.56 ± 2.46	191.44 ± 127.32	142.11 ± 2.42	364.67 ± 17.60	113.33 ± 2.06	206.70 ± 137.55
4. Velocity (m/sec)	1.59 ± 0.28	1.85 ± 0.03	0.73 ± 0.05	1.39 ± 0.58	1.76 ± 0.10	1.88 ± 0.11	1.02 ± 0.03	1.55 ± 0.47
5. Total Solids (mg/l)	552.70 ± 47.92	3846.44 ± 150.21	156.44 ± 2.92	1518.53 ± 2025.75	604.56 ± 49.73	4063.67 ± 152.44	179.22 ± 13.25	1615.81 ± 2130.54
6. Total Dissolved Solids (mg/l)	220.13 ± 6.94	852.56 ± 22.34	48.33 ± 4.50	373.67 ± 423.53	287.33 ± 41.08	1012.89 ± 36.89	34.00 ± 8.59	444.74 ± 508.07
7. Total Suspended Solids (mg/l)	332.57 ± 45.63	2993.89 ± 147.08	108.11 ± 4.23	1144.86 ± 1605.24	313.67 ± 29.16	3050.78 ± 129.58	145.22 ± 7.34	1169.89 ± 1631.07
8. pH	7.30 ± 0.27	7.57 ± 0.15	7.07 ± 0.10	7.32 ± 0.25	7.28 ± 0.13	7.33 ± 0.21	7.00 ± 0.23	7.20 ± 0.18
9. Free Carbon Dioxide (mg/l)	2.21 ± 0.29	5.01 ± 0.19	1.75 ± 0.07	2.99 ± 1.77	1.97 ± 0.17	4.80 ± 0.45	1.64 ± 0.07	2.80 ± 1.74
10. Dissolved Oxygen (mg/l)	9.52 ± 0.89	7.38 ± 0.34	11.74 ± 0.43	9.55 ± 2.18	10.02 ± 0.87	7.38 ± 0.33	11.28 ± 0.17	9.56 ± 1.99

Table 4A.7. (Contd.)

11. B.O.D. (mg/l)	2.36 ± 0.27	3.02 ± 0.56	1.39 ± 0.10	2.26 ± 0.82	2.37 ± 0.13	2.77 ± 0.49	1.72 ± 0.55	2.29 ± 0.53
12. C.O.D. (mg/l)	9.09 ± 1.02	5.36 ± 2.17	11.62 ± 0.40	8.69 ± 3.15	9.14 ± 0.48	6.63 ± 1.89	10.68 ± 2.10	8.82 ± 2.04
13 Total Alkalinity (mg/l)	66.67 ± 2.50	48.06 ± 0.46	35.33 ± 0.50	50.02 ± 15.76	69.11 ± 1.05	48.44 ± 1.13	38.22 ± 1.09	51.93 ± 15.74
14 Chloride (mg/l)	2.96 ± 0.03	11.58 ± 0.26	1.97 ± 0.04	5.51 ± 5.29	2.70 ± 0.19	11.36 ± 0.12	1.85 ± 0.24	5.31 ± 5.26
15. Total Phosphate (mg/l)	0.02 ± 0.00	0.11 ± 0.01	0.07 ± 0.01	0.07 ± 0.04	0.02 ± 0.00	0.14 ± 0.01	0.08 ± 0.01	0.08 ± 0.06
16. Total Organic Carbon (mg/l)	3.37 ± 0.27	5.18 ± 0.04	4.14 ± 0.03	4.23 ± 0.91	3.33 ± 0.21	5.27 ± 0.13	4.68 ± 0.18	4.43 ± 1.00
17 Hardness (mg/l)	83.78 ± 1.30	107.20 ± 2.27	96.00 ± 0.71	95.66 ± 11.71	88.78 ± 1.20	113.83 ± 2.96	90.11 ± 9.56	97.57 ± 14.10
18. Ions								
(*a*) Sodium (mg/l)	8.9227 ± 0.02	23.1768 ± 0.03	11.3687 ± 0.02	14.4894 ± 7.62	10.2147 ± 0.01	25.8996 ± 0.02	13.0665 ± 0.02	16.3936 ± 8.36
(*b*) Potassium (mg/l)	1.1842 ± 0.02	2.5928 ± 0.02	1.6139 ± 0.02	1.7970 ± 0.72	1.1842 ± 0.01	2.6608 ± 0.01	1.6042 ± 0.02	1.8164 ± 0.76
(*c*) Sulphate (mg/l)	18.2000 ± 0.02	23.1600 ± 0.02	19.5000 ± 0.01	20.2867 ± 2.57	15.5400 ± 0.02	23.1600 ± 0.02	17.9800 ± 0.03	19.8933 ± 2.84
(*d*) Nitrate (mg/l)	0.0150 ± 0.01	0.0700 ± 0.01	0.0300 ± 0.01	0.0383 ± 0.03	0.0130 ± 0.01	0.0700 ± 0.01	0.0160 ± 0.01	0.0330 ± 0.03

± = Standard Deviation.

Table 4A.8. Seasonal Variation in Heavy metals of River Ganga at Sampling Station D (1998-2000)

Seasons Parameters	1998 to 1999				1999 to 2000			
	Summer	Monsoon	Winter	Average	Summer	Monsoon	Winter	Average
Lead [Pb] (mg/l)	0.0686 ± 0.0029	0.0784 ± 0.0012	0.0643 ± 0.0013	0.0704 ± 0.0072	0.0727 ± 0.0040	0.0798 ± 0.0015	0.0610 ± 0.0013	0.0712 ± 0.0095
Copper [Cu] (mg/l)	0.0169 ± 0.0004	0.0183 ± 0.0002	0.0176 ± 0.0003	0.0176 ± 0.0007	0.0181 ± 0.0001	0.0185 ± 0.0003	0.0172 ± 0.0002	0.0179 ± 0.0006
Chromium [Cr] (mg/l)	0.1123 ± 0.1539	0.0294 ± 0.0016	0.0310 ± 0.0011	0.0576 ± 0.0474	0.1397 ± 0.1960	0.0351 ± 0.0007	0.0269 ± 0.0001	0.0672 ± 0.0629
Cadmium [Cd] (mg/l)	0.0031 ± 0.0002	0.0034 ± 0.0003	0.0033 ± 0.0003	0.0032 ± 0.0001	0.0036 ± 0.0004	0.0034 ± 0.0004	0.0026 ± 0.0002	0.0032 ± 0.0005
Zinc [Zn] (mg/l)	3.2250 ± 0.0218	3.2551 ± 0.0336	3.1802 ± 0.0286	3.2201 ± 0.0377	3.2097 ± 0.0090	3.2619 ± 0.0018	3.2019 ± 0.0009	3.2245 ± 0.0326

± = Standard Deviation.

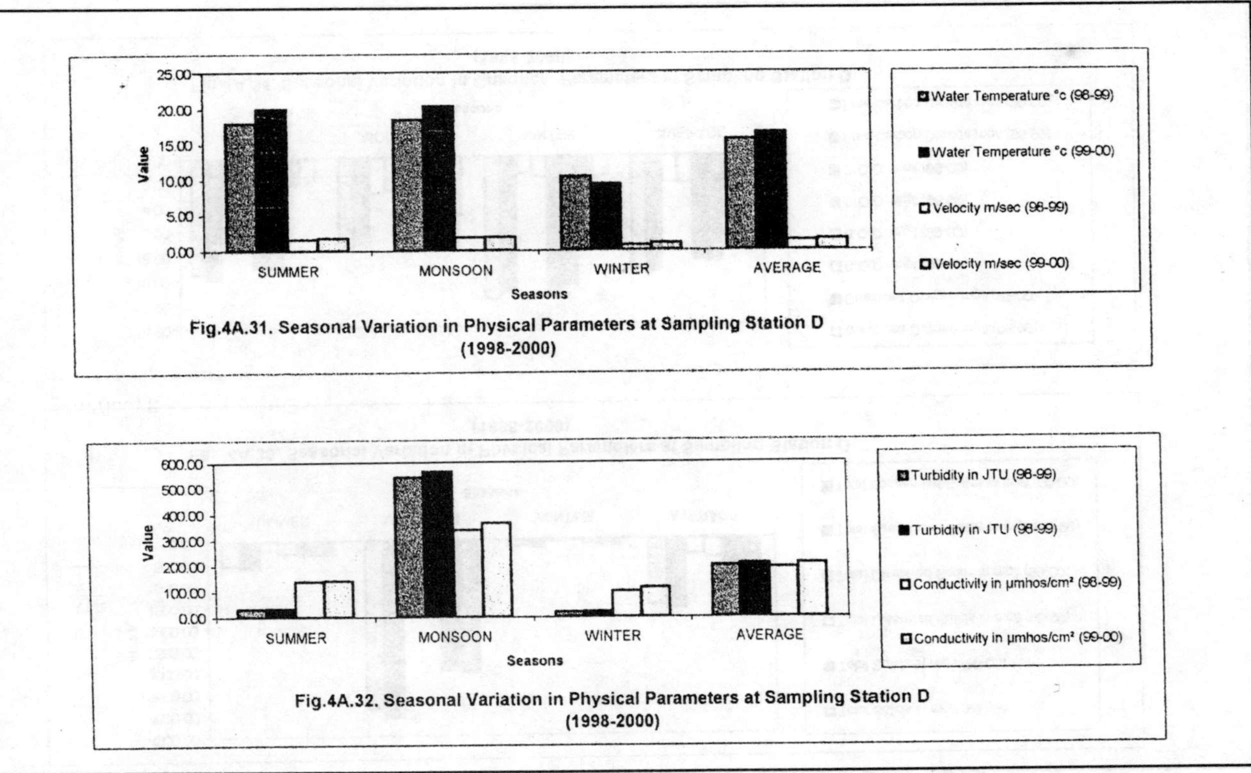

Fig.4A.31. Seasonal Variation in Physical Parameters at Sampling Station D (1998-2000)

Fig.4A.32. Seasonal Variation in Physical Parameters at Sampling Station D (1998-2000)

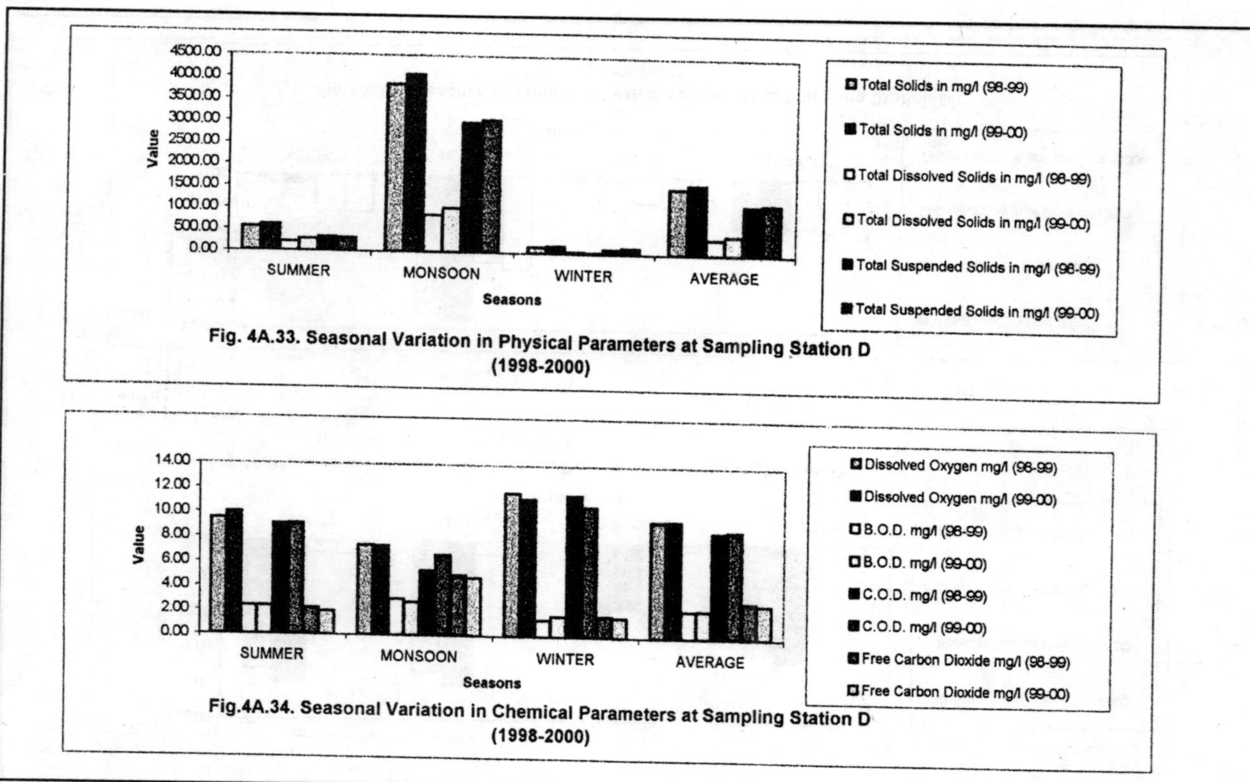

Fig. 4A.33. Seasonal Variation in Physical Parameters at Sampling Station D (1998-2000)

Fig.4A.34. Seasonal Variation in Chemical Parameters at Sampling Station D (1998-2000)

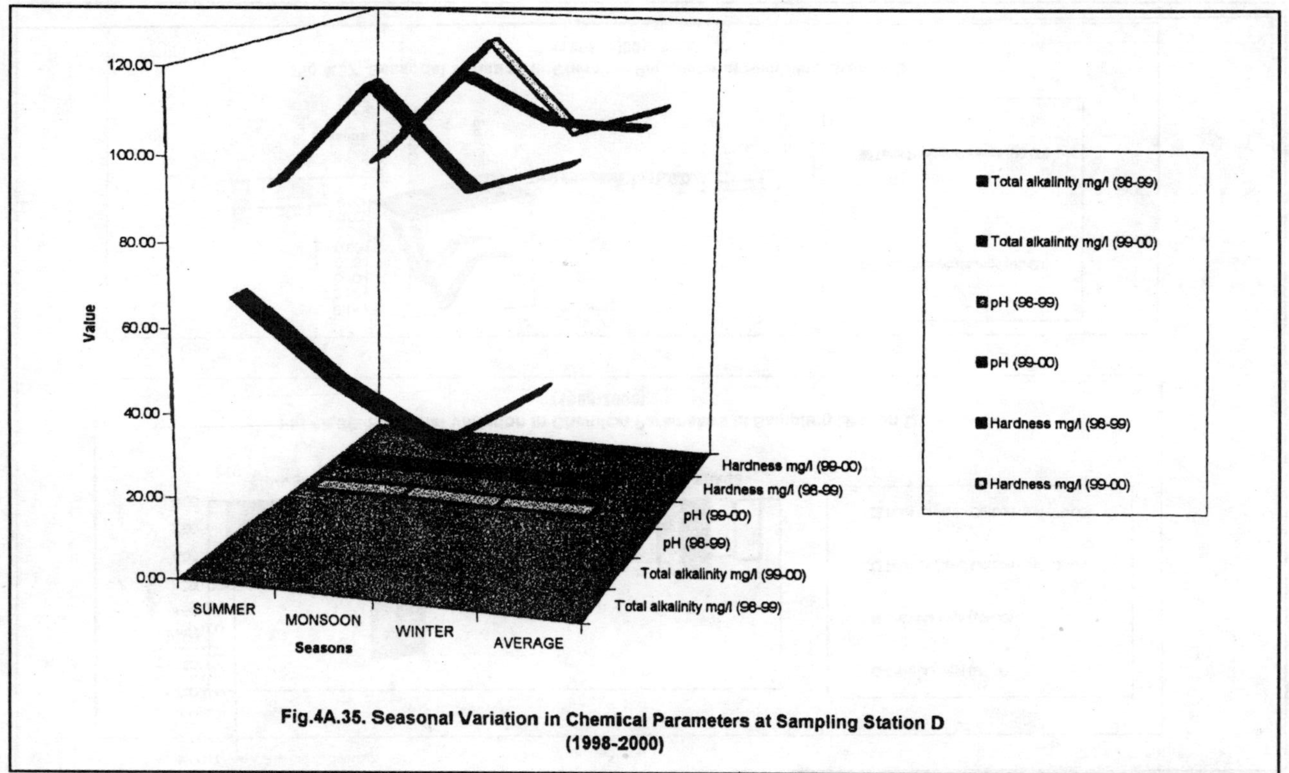

Fig.4A.35. Seasonal Variation in Chemical Parameters at Sampling Station D (1998-2000)

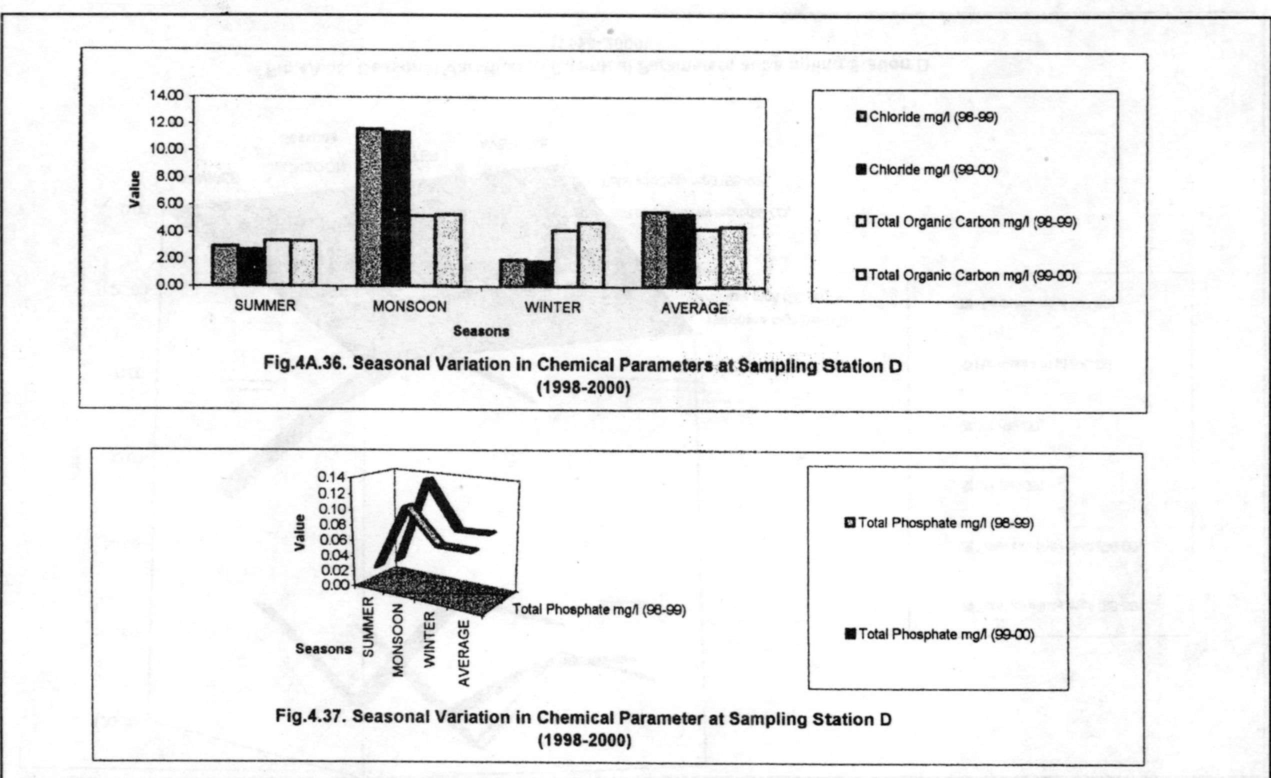

Fig.4A.36. Seasonal Variation in Chemical Parameters at Sampling Station D (1998-2000)

Fig.4.37. Seasonal Variation in Chemical Parameter at Sampling Station D (1998-2000)

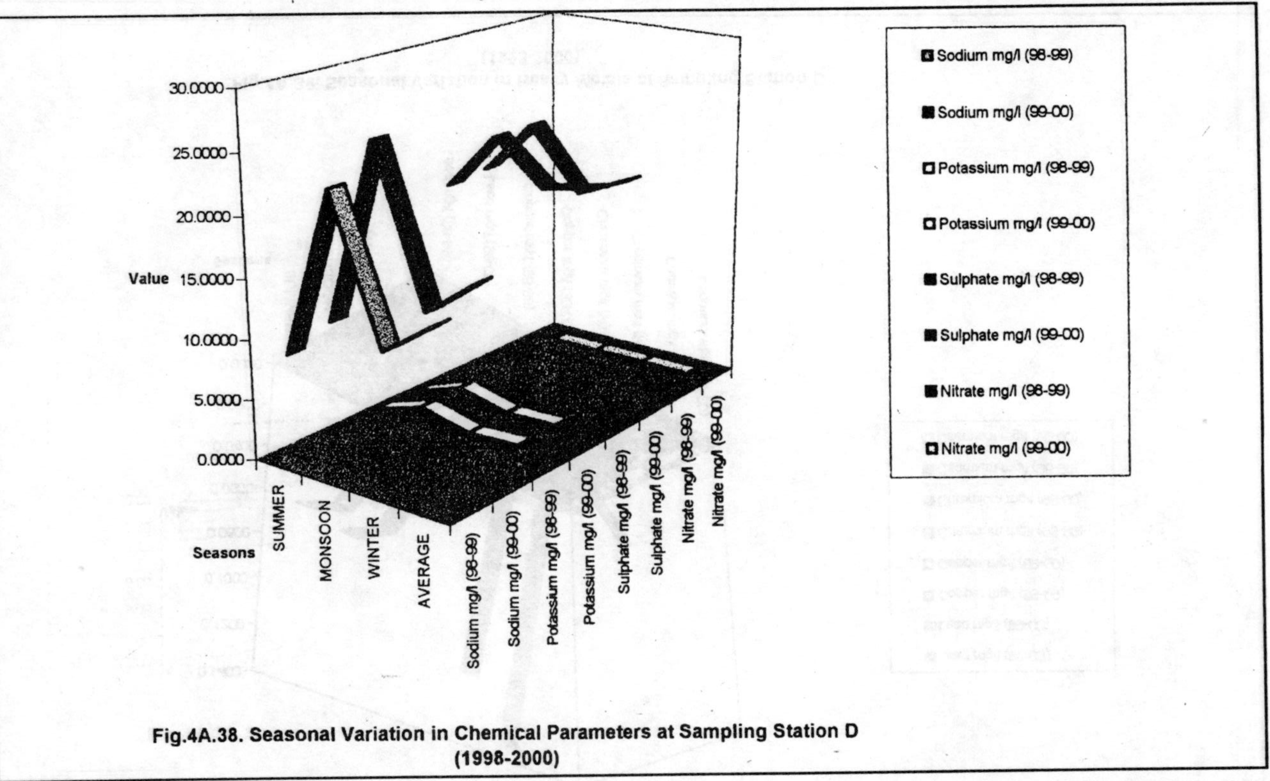

Fig.4A.38. Seasonal Variation in Chemical Parameters at Sampling Station D (1998-2000)

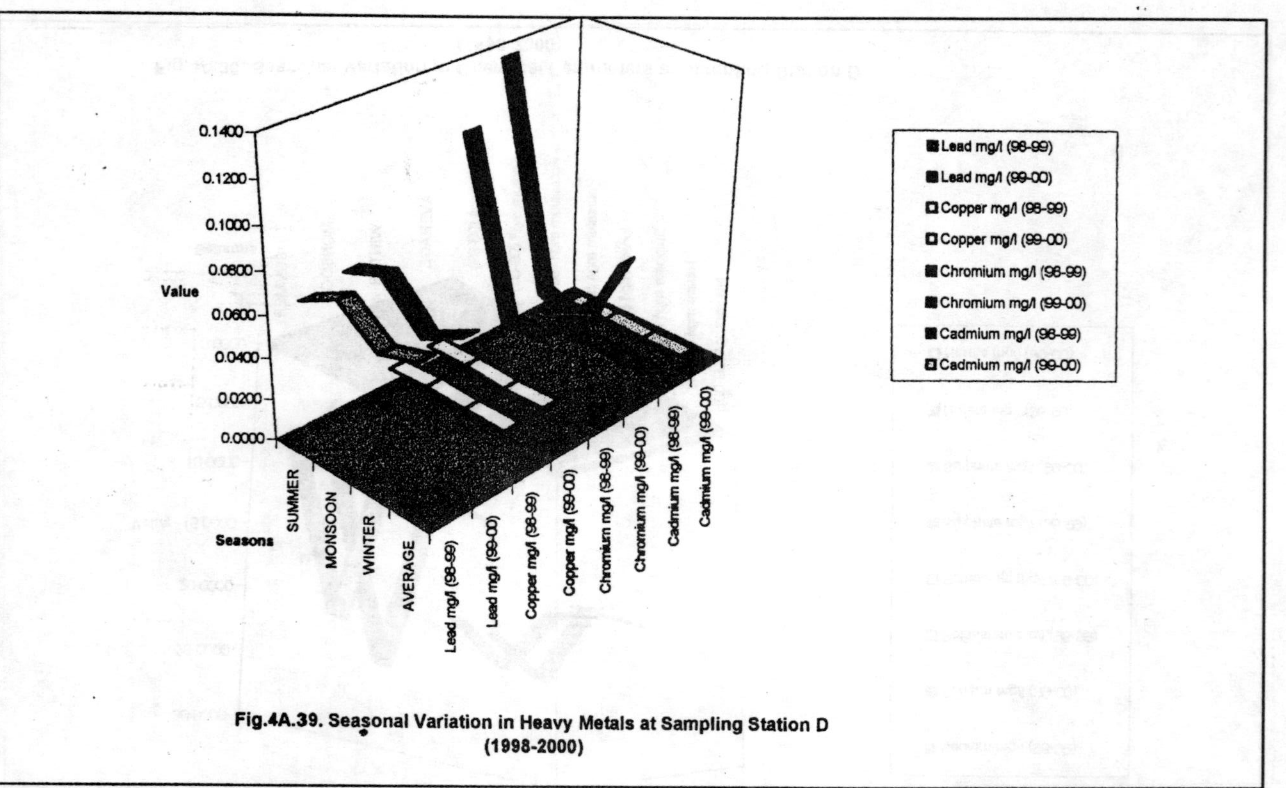

Fig.4A.39. Seasonal Variation in Heavy Metals at Sampling Station D (1998-2000)

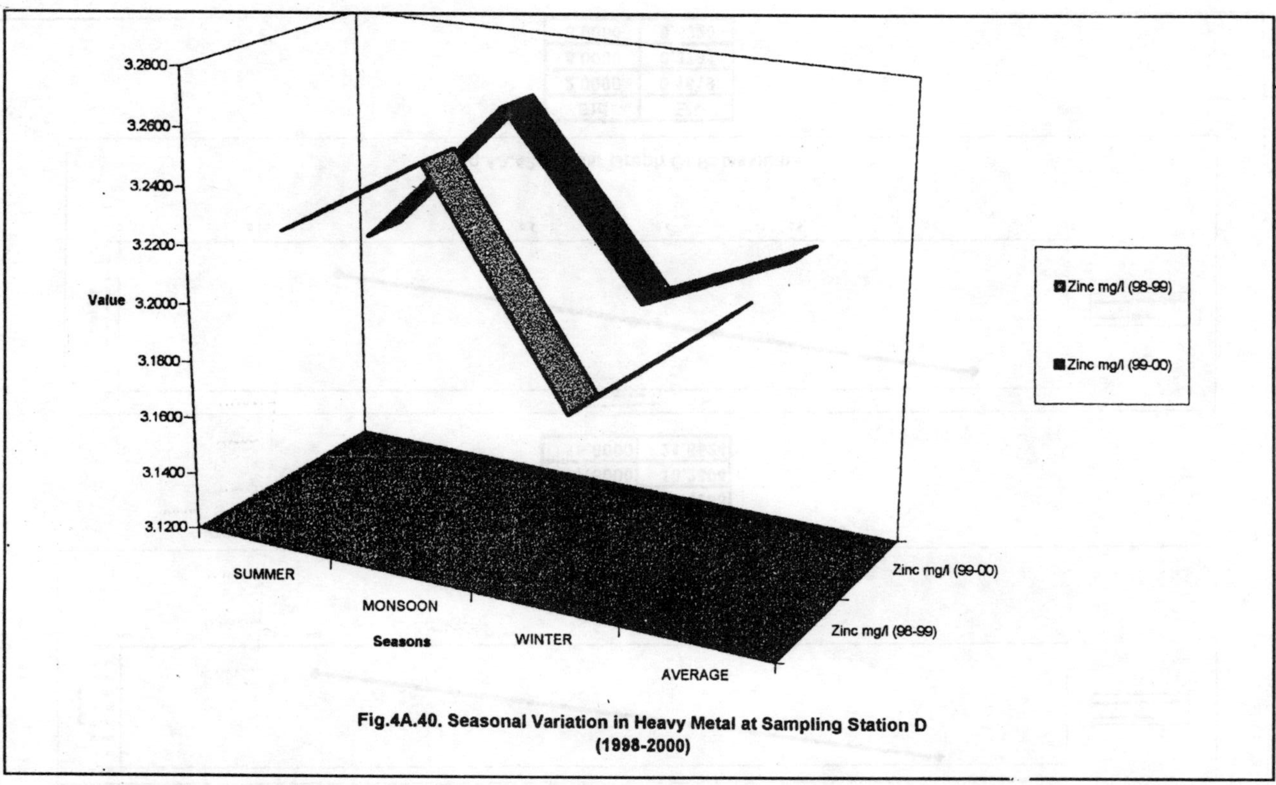

Fig.4A.40. Seasonal Variation in Heavy Metal at Sampling Station D (1998-2000)

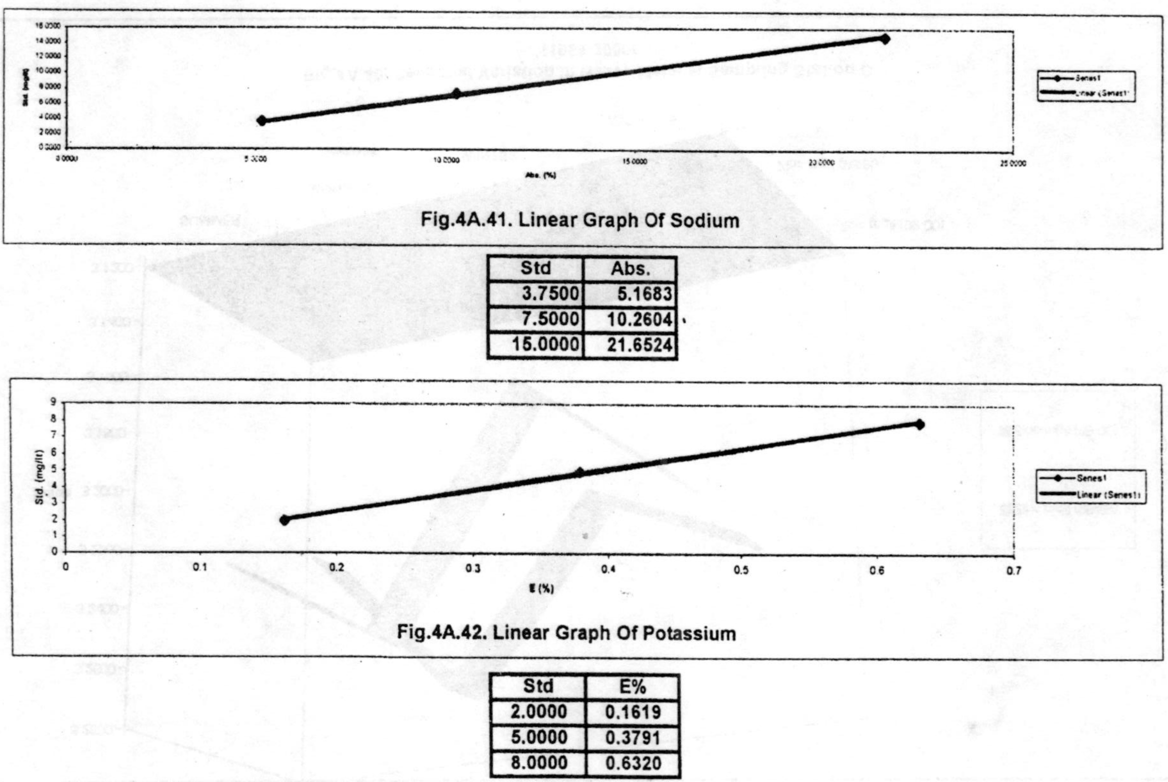

Fig.4A.41. Linear Graph Of Sodium

Std	Abs.
3.7500	5.1683
7.5000	10.2604
15.0000	21.6524

Fig.4A.42. Linear Graph Of Potassium

Std	E%
2.0000	0.1619
5.0000	0.3791
8.0000	0.6320

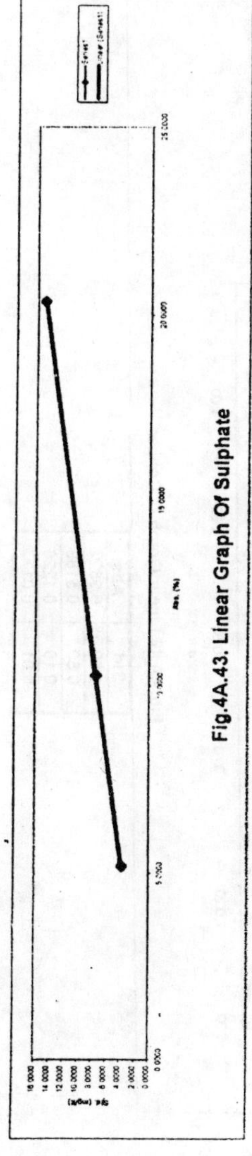

Fig.4A.43. Linear Graph Of Sulphate

Std	Abs.
3.7500	5.1802
7.5000	10.2809
15.0000	20.3613

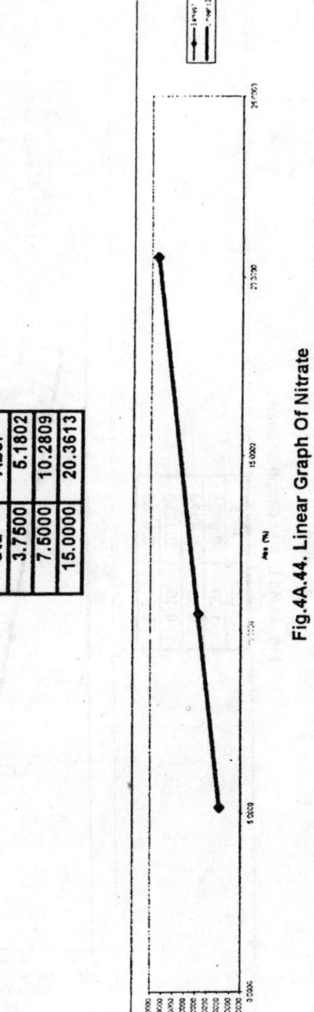

Fig.4A.44. Linear Graph Of Nitrate

Std	Abs.
3.7500	5.1594
7.5000	10.5484
15.0000	20.5589

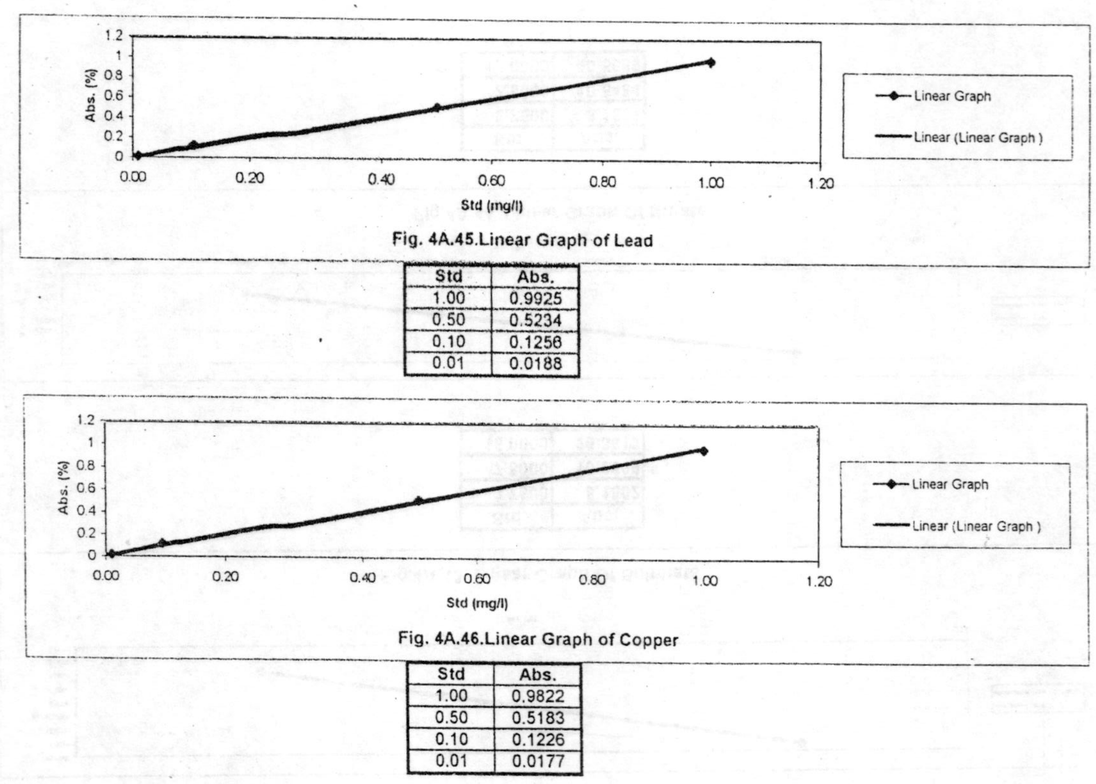

Fig. 4A.45. Linear Graph of Lead

Std	Abs.
1.00	0.9925
0.50	0.5234
0.10	0.1256
0.01	0.0188

Fig. 4A.46. Linear Graph of Copper

Std	Abs.
1.00	0.9822
0.50	0.5183
0.10	0.1226
0.01	0.0177

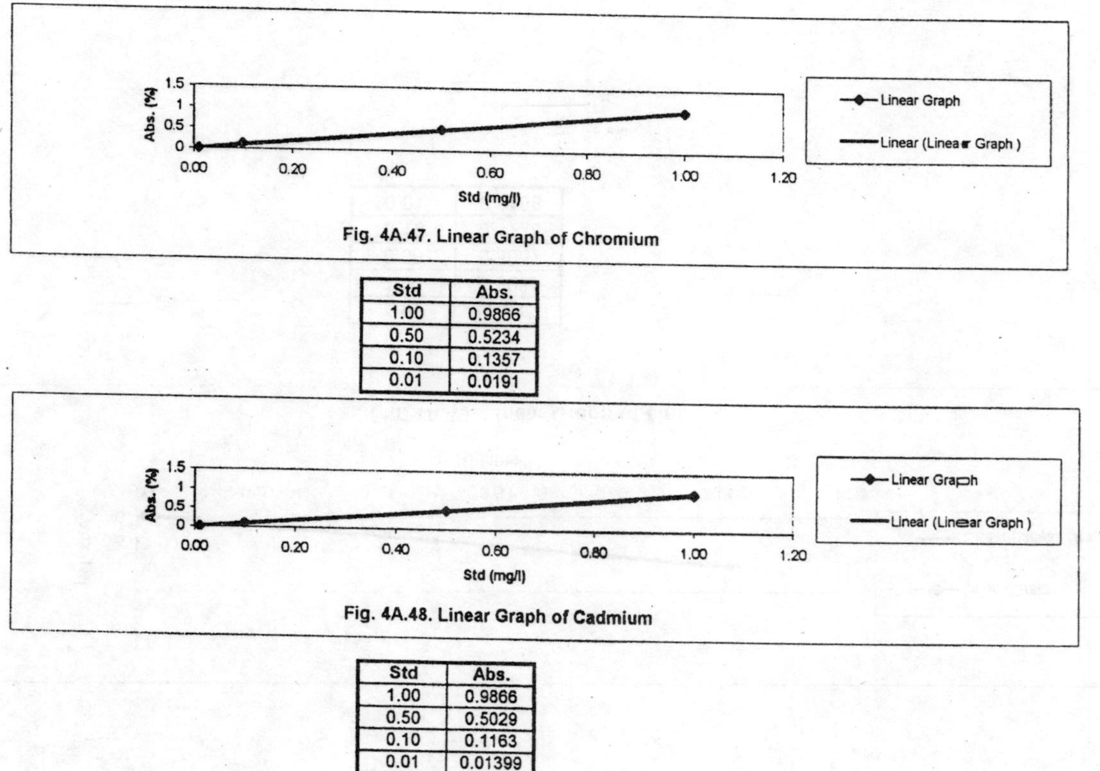

Fig. 4A.47. Linear Graph of Chromium

Std	Abs.
1.00	0.9866
0.50	0.5234
0.10	0.1357
0.01	0.0191

Fig. 4A.48. Linear Graph of Cadmium

Std	Abs.
1.00	0.9866
0.50	0.5029
0.10	0.1163
0.01	0.01399

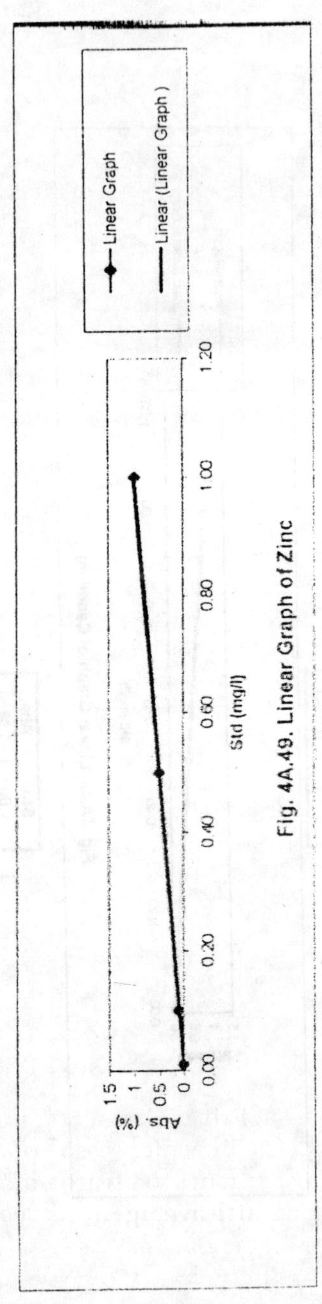

Fig. 4A.49. Linear Graph of Zinc

Std	Abs.
1.00	0.9967
0.50	0.5067
0.10	0.1228
0.01	0.0168

3. Conductivity

The conductivity of water at Pul-Jatwara was observed to be least in winter season of 1998-1999, 99.56 μmhos/cm^2 ±2.46, which goes highest up to 364.67 μmhos/cm^2 ±17.60 in monsoon season of 1999-2000. The annual average observed for first year (1998-1999) was 191.44 μmhos/cm^2 ±127.32 and second year (1999-2000) was 206.70 μmhos/cm^2 ±137.55.

4. Velocity

The velocity of this sampling station at river Ganga was noted to be minimum in winter season of 1998-1999, 0.73 m/sec ±0.05, which goes maximum up to 1.88 m/sec ±0.11 in monsoon season of 1999-2000. The annual average calculated for first year (1998-1999) was 1.39 m/sec ±0.58 and second year (1999-2000) was 1.55 m/sec ±0.47.

5. Total Solids

The total solids of river Ganga at this spot was recorded to be lowest in 1998-1999's winter season, 156.44 mg/l ±2.92, which goes highest up to 4063.67 mg/l ±152.44 in 1999-2000's monsoon season. The annual mean recorded for first year (1998-1999), 1518.53 mg/l ±2025.75 and second year (1999-2000) was 1615.81 mg/l ±2130.54.

6. Total Dissolve Solids

At Pul-Jatwara the total dissolved solids of river Ganga was recorded to be minimum and maximum both in 1999-2000 i.e., in winter season 34 mg/l ±8.59, and in monsoon season 1012.89 mg/l ±36.89 respectively. The annual average recorded for first year (1998-1999) 373.67 mg/l ±423.53 and second year (1999-2000) 444.74 mg/l ±508.07.

7. Total Suspended Solids

The total suspended solids of river Ganga at this sampling station were observed to be lowest in winter season of 1998-1999 (108.11 mg/l ±4.23) and highest in monsoon season of 1999-2000 (3050.78 mg/l ±129.58). The annual average recorded for first year (1998-1999) was 1144.86 mg/l ±1605.24 and second year (1999-2000) 1169.89 mg/l ±1631.07.

8. pH

At this sampling station the pH of river Ganga was noted to be least in winter season 7 mg/l ±0.23 (1999-2000) and optimum in monsoon season of 1998-1999 (7.57 mg/l ±0.15). The annual average recorded for first year (1998-1999) was calculated 7.32 ±0.25 and second year (1999-2000) was 7.20 ±0.18.

9. Free Carbon Dioxide

The free carbon dioxide of river Ganga at this sampling station was observed to be lowest in winter season (1999-2000), 1.64 mg/l ±0.07, which goes highest up to 5.01 mg/l ±0.19 in monsoon season (1998-1999). The annual mean recorded for first year (1998-1999) 2.99 mg/l ±1.77 and second year (1999-2000) 2.80 mg/l ±1.74.

10. Dissolved Oxygen

The dissolved oxygen of river Ganga at this sampling station was recorded to be minimum in monsoon season 7.38 mg/l ±0.33, which goes maximum up to 11.74 mg/l ±0.43 in winter season (1998-1999). The annual average observed for first year (1998-1999), 9.55 mg/l ±2.18 and second year (1999-2000), was 9.56 mg/l ±1.99.

11. Biological Oxygen Demand (BOD)

In the same year (1998-1999) the biological oxygen demand (BOD) of river Ganga at this spot was noted to be lowest in winter season 1.39 mg/l ±0.10, which goes highest up to 3.02 mg/l ±0.56 in monsoon season. The annual average recorded for first year (1998-1999) 2.26 mg/l ±0.82 and second year (1999-2000) 2.29 mg/l ±0.53.

12. Chemical Oxygen Demand (COD)

The chemical oxygen demand (COD) of river Ganga at this sampling station was observed to be least in monsoon season (1998-1999) 5.36 mg/l ±2.17, which goes highest up to 11.62 mg/l ±0.40 in winter season (1998-1999). The annual mean calculated for first year (1998-1999) 8.69 mg/l ±3.15 and second year (1999-2000) 8.82 mg/l ±2.04.

13. Total Alkalinity

The total alkalinity of water at Pul-Jatwara was observed to be minimum in winter season (1998-1999), 35-33 mg/l ±0.50, which goes highest up to 69.11 mg/l ±1.05 in summer season (1999-2000). The annual average recorded for first year (1998-1999) was 50.02 mg/l ±15.76 and second year (1999-2000) was 51.93 mg/l ±15.74.

14. Chloride

The chloride in the water of river Ganga at Pul-Jatwara this spot was observed to be lowest in winter season (1999-2000), 1.85 mg/l ±0.24, which goes highest up to 11.58 mg/l ±0.26 in monsoon season (1998-1999). The annual average recorded for first year (1998-1999) was 5.51 mg/l ±5.29 and second year (1999-2000) was 5.31 mg/l ±5.26.

15. Total Phosphate

At this sampling station the total phosphate of river Ganga was observed to be minimum in summer season of 1998-1999 and 1999-2000, 0.02 mg/l ±0.00, which goes maximum up to 0.14 mg/l ±0.01 in monsoon season (1999-2000). The annual average observed for first year (1998-1999) was 0.07 mg/l ±0.04 and second year (1999-2000) was 0.08 mg/l ±0.06.

16. Total Organic Carbon (TOC)

The total organic carbon (TOC) at this spot of river Ganga was recorded to be lowest in summer season (1999-2000), 3.33 mg/l ±0.21, which goes maximum up to 5.27 mg/l ±0.13 in monsoon season (1999-2000). The annual mean noted for first year (1998-1999) was mg/l ±0.91 and second year (1999-2000) was 4.43 mg/l ±1.00.

17. Hardness

The hardness at this sampling station of river Ganga was observed to be least in summer season (1998-1999), 83.78 mg/l ±1.30, which goes maximum up to 113.83 mg/l ±2.96 in monsoon season (1999-2000). The annual average observed for first year (1998-1999) was 95.66 mg/l ±11.71 and second year (1999-2000) was 97.57 mg/l ±14.10.

18. Ions

(a) Sodium

The sodium at this spot of river Ganga was recorded to be minimum in summer season (1998-1999), 8.9227 mg/l ±0.02, which goes maximum up to 25.8996 mg/l ±0.02 in monsoon season (1999-2000). The annual average was found 14.4894 mg/l ±7.62 for first year (1998-1999) and 16.3936 mg/l ±8.36 for second year (1999-2000).

(b) *Potassium*

At this sampling station the potassium of river Ganga was observed to be lowest in winter season (1999-2000), 1.6042 mg/l ±0.02, which goes highest up to 2.6608 mg/l ±0.01 in monsoon season (1999-2000). The annual mean calculated for first year (1998-1999) was 1.7970 mg/l ±0.72 and second year (1999-2000) was 1.8164 mg/l ±0.76.

(c) *Sulphate*

At this sampling station the sulphate of river Ganga was recorded to be minimum in winter season (1999-2000), 17.9800 mg/l ±0.03, which goes maximum up to 23.16000 mg/l ±0.02 in monsoon season of both years. The annual average observed for first year (1998-1999) was 20.2867 mg/l ±2.57 and second year (1999-2000) was 19.8933 mg/l ±2.84.

(d) *Nitrate*

The nitrate in water at Paul-Jatwara was observed to be lowest in summer season (1999-2000), 0.0130 mg/l ±0.01, which goes highest up to 0.07 mg/l ±0.01 in monsoon season of 1998-1999 and 1999-2000. The annual average recorded for first year (1998-1999) was 0.0383 mg/l ±0.03 and second year (1999-2000) was 0.0330 mg/l ±0.03.

19. Heavy Metals

(a) *Lead (Pb)*

The lead at this sampling station of river Ganga was recorded to be minimum in winter season (1999-2000), 0.0610 mg/l ±0.0013, which goes maximum up to 0.0798 mg/l ±0.0015 in monsoon season (1999-2000). The annual average for first year (1998-1999) was 0.0704 mg/l ±0.0072 and second year (1999-2000) was 0.0712 mg/l ±0.0095.

(b) *Copper (Cu)*

At this spot the copper was recorded to be lowest in winter season (1999-2000), 0.0172 mg/l ±0.0002, which goes highest up to 0.0185 mg/l ±0.0003 in monsoon season (1999-2000). The annual average observed for first year (1998-1999) was 0.0176 mg/l ±0.0007 and second year (1999-2000) was 0.0179 mg/l ±0.0006.

(c) *Chromium (Cr)*

At this sampling station the chromium of river Ganga was recorded to be least in winter season (1999-2000), 0.0269 mg/l ±0.0001, which goes optimum up to 0.0351 mg/l ±0.0007 in monsoon season (1999-2000). The annual average calculated for first year (1998-1999) was 0.0576 mg/l ±0.0474 and second year (1999-2000) was 0.0672 mg/l ±0.0629.

(d) *Cadmium (Cd)*

The cadmium at this sampling station of river Ganga was observed to be lowest in winter season (1999-2000), 0.0026 mg/l ±0.0002, which goes highest up to 0.0036 mg/l ±0.0004 in summer season (1999-2000). The annual average observed for first year (1998-1999) was 0.0032 mg/l ±0.0001 and second year (1999-2000) was 0.0032 mg/l ±0.0005.

(e) *Zinc (Zn)*

At this spot the zinc was recorded to be minimum in winter season (1998-1999), 3.1802 mg/l ±0.0286, which goes maximum up to 3.2619 mg/l ±0.0018 in monsoon season (1999-2000). The annual mean calculated for first year (1998-1999) was 3.2201 mg/l ±0.0377 and second year (1999-2000) was 3.2245 mg/l ±0.326.

B. MICROBIOLOGICAL PARAMETERS

Studies on the microbiological parameters of the river Ganga include the primary producer like phytoplankton, primary consumer like zooplankton, and bacterial population. The biological and microbiological parameters of the river Ganga are discussed in the following pages.

Among zooplankton – **Protozoa, Rotifera, Cladocera,** and **Copepoda** were found in this study. Further, in phylum: **Protozoa** – genus: *Arcella* and *Amoeba;* in phylum: **Rotifera** – genus: *Keratella, Philodima* and *Notholca;* in phylum: **Cladocera** –

genus: *Daphnia* and *Bosminar;* in phylum: **Copepoda** – genus: *Cyclops* were found.

In phytoplankton – **Bacillariophyceae, Chlorophyceae** and **Myxophyceae** were observed in this study. Among class **Bacillariophyceae** (Type *Diatoms)* – *Achnanthes, Amphora, Cocconeis, Cymbella, Cyclotella, Diatoms, Fragilaria, Frustulia, Gomphonema, Navicula, Nitzschia, Pinnularia, Phacus, Stauroneis, Surirella, Synedra, Tabellaria* were found. In class **Chlorophyceae** – *Closterium, Cosmarium, Chaetophora, Chlorella, Cladophora, Oedogonium, Schizogonium, Spirogyra, Ulothrix, Vaucheria, Zygnema* were found and in class **Myxophyceae** (Blue green algae) – *Anacystis, Anabaena, Lyngbya, Oscillatoria* were found at Hardwar.

In bacteriology – Most Probable Number (MPN) and Standard Plate Count (SPC) (Plate 4.1-Plate 4.4) examination were carried out. Among bacteria in this study at Hardwar – *Escherichia coli, Pseudomonas aeruginosa, Bacillus subtilis, Proteus mirabilis, Clostridium perfringes, Streptococcus facecalis, Aerobacter aerogens, Salmonella typhii, Staphylococcus aureus, Sporolactobacillus sp., Klebssilla pneumoniae, Staphylococcus epidermidis, Micrococcus luteus* were recorded.

Description of Isolated Bacterial Culture of River Ganga at Hardwar

Following is brief description of bacteria cultures isolated and identified during this study from river Ganga at Hardwar (Plates 4.5-4.6).

Escherichia coli

A genus of aerobic, facultatively anaerobic bacteria containing short, motile or nonmotile, gram-negative rods. Motile cells are peritrichous. Glucose and lactose are fermented with the production of acid and gas. These organisms are found in feces; some are pathogenic to man, causing enteritis, peritonitis, cystitis, etc. It is the type genus of the family Enterobacteriaceae.

Pseudomonas aeruginosa

A genus of motile, polar flagellate, nonsporeforming, strictly aerobic bacteria (family Pseudomonadaceae) containing

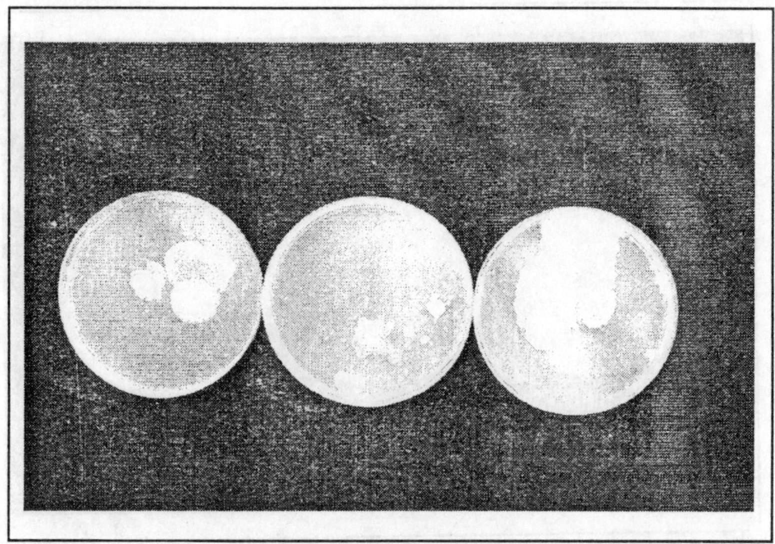

Plate 4.1. Standard Plate Count (SPC) of Har-ki-Pauri

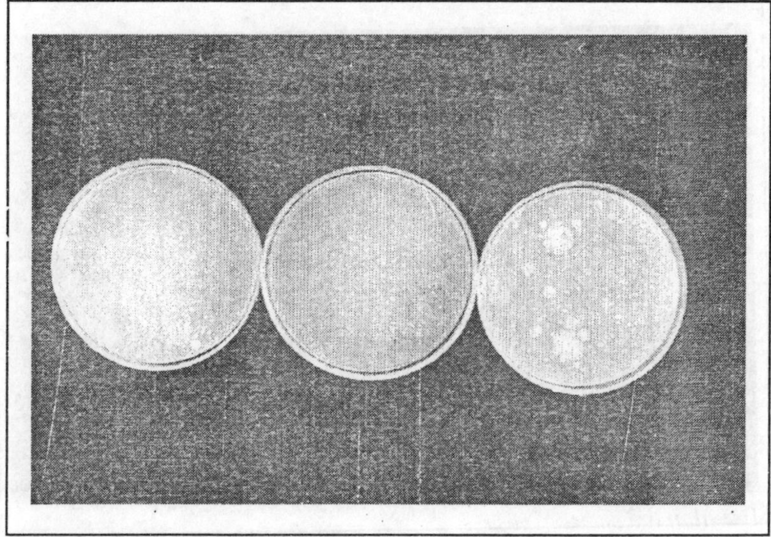

Plate 4.2. Standard Plate Count (SPC) of Mayapuri Ghat

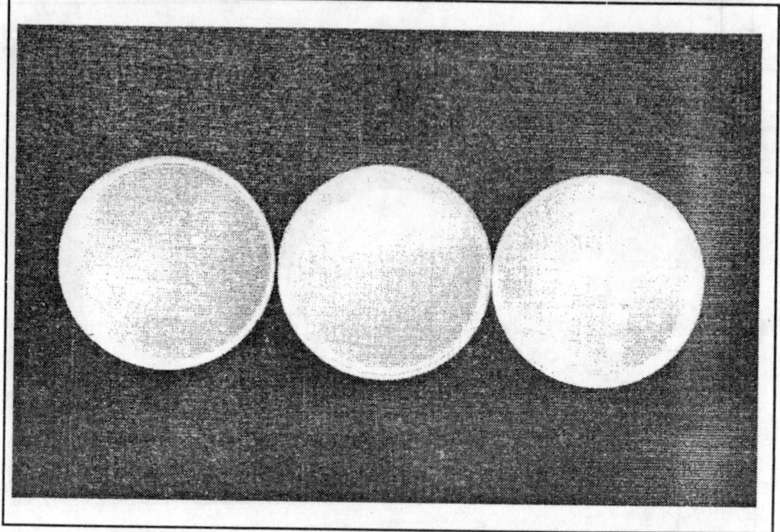

Plate 4.3. Standard Plate Count (SPC) Muneshwar Ghat

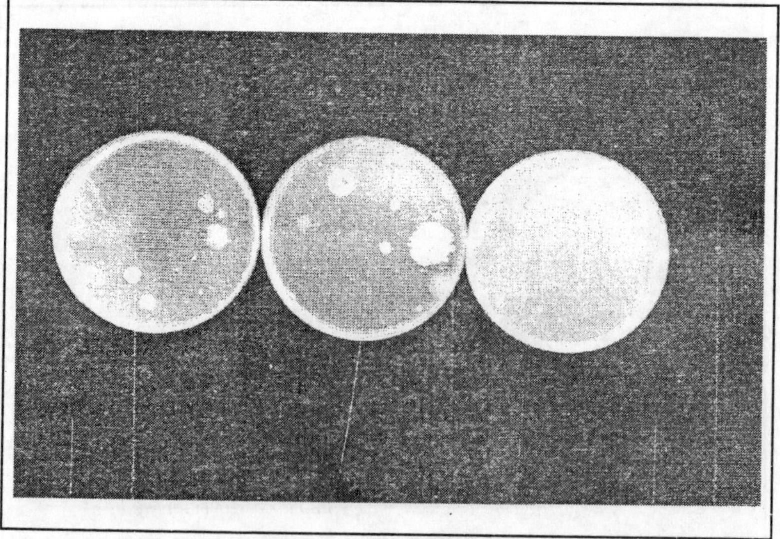

Plate 4.4. Standard Plate Count (SPC) Pul-Jatwara

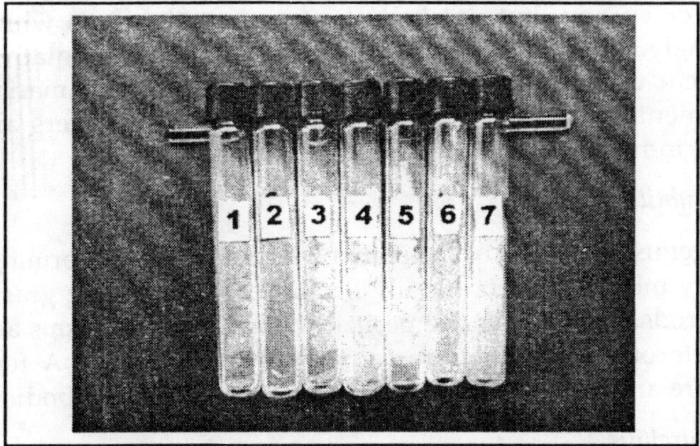

Plate 4.5. Isolated Bacterial Cultures in Slant 1-7

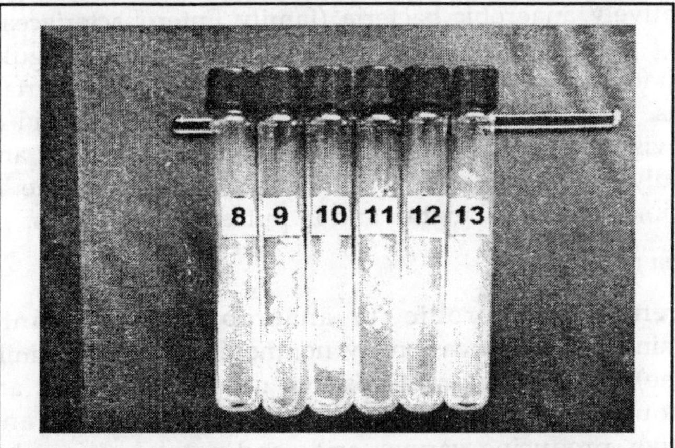

Plate 4.6. Isolated Bacterial Cultures in Slant 8-13

Isolated Bacteria in slants:

1. *Escherichia coli*
2. *Pseudomonas aeruginosa*
3. *Bacillus subtilis*
4. *Proteus mirabilis*
5. *Clostridium perfringes*
6. *Streptococcus facecalis*
7. *Aerobacter aerogens*
8. *Salmonell typhii*
9. *Staphylococcus anrens*
10. *Sporolactobacillus sp.*
11. *Klebssilla pneumoniae*
12. *Staphlococcus epidermidis*
13. *Micrococcus luteus*

straight or curved, but not helical, gram-negative rods which occur singly. The metabolism is respiratory, never fermentative. They occur commonly in soil and in fresh water and marine environments. Some species are plant pathogens. Others are involved in human infections.

Bacillus subtilis

A genus of aerobic or facultatively anaerobic, sporeforming, ordinarily motile bacteria (family Bacillaceae) containing gram-positive rods. Motile cells are peritrichous. These organisms are chemoheterotrophic. They are found primarily in soil. A few species are animal pathogens; some species produce antibodies.

Proteus mirabilis

A genus of motile, peritrichous, nonsporeforming, aerobic to facultatively anaerobic bacteria (family Enterobacteriaceae) containing gram-negative rods; coccoid forms, large irregular involution forms, filaments, and spheroplasts occur under certain conditions. The metabolism is fermentative, producing acid or acid and visible gas from glucose; lactose is not fermented, and they rapidly decompose urea and deaminate phenylalanine. *P.* occurs primarily in fecal matter and in putrefying materials.

Clostridium perfringes

A genus of anaerobic (or anaerobic, aerotolerant), sporeforming, motile (occasionally nonmotile) bacteria (family Bacillaceae) containing gram-positive rods; motile calls are peritrichous. Many of the species are saccharolytic and fermentative, producing various acids and gases and variable amounts a neutral products; other species are proteolytic, some attacking proteins with putrefaction or more complete proteolysis. Some species fix free nitrogen. Exotoxins are sometimes produced by these organisms. They may cause disease in man and other animals. They are generally found in soil and in the intestinal tract of man and other animals.

Streptococcus facecalis

A genus of nonmotile, nonsporeforming, aerobic to facultatively anaerobic bacteria (family Lactobacillaceae) containing gram-positive, spherical or ovoid cells which occur in pairs or short or long chains. Dextrorotatory lactic acid is the main product of carbohydrate fermentation. These organisms occur regularly in the mouth and intestines of humans and other animals, in dairy and other food products, and in fermenting plant juices.

Aerobacter aerogens

The type species is *A. aerogenes.* Motile organisms previously placed in this species are now placed in *Enterobacter aerogenes;* the nonmotile organisms have been transferreo to *Klebsiella pneumoniae.*

Salmonella typhii

A genus of aerobic to facultatively anaerobic bacteria (family Enterobacteriaceae) containing Gram-negative rods that are either motile or nonmotile; motile cells are peritrichous. These organisms do not liquefy gelatin or produce indole and very in their production of hydrogen sulfide; they utilize citrate as a sole source of carbon; their metabolism is fermentative, producing acid and usually gas from glucose, but they do not attack lactose; most are aerogenic, but *S. typhii* never produces gas; they are pathogenic for humans and other animals.

Streptococcus facecalis/Staphylococcus epidermidis

A genus of nonmotile, nonsporeforming, aerobic to facultatively anaerobic bacteria (family Micrococcaceae) containing gram-positive, spherical cells, 0.5 to 1.5 mum in diameter, which divide in more than one plane to form irregular clusters. These organisms are chemoorganotrophic, and their metabolism is respiratory and fermentative. Under anaerobic conditions, lactic acid is produced from glucose; under aerobic conditions, acetic acid and small amounts of CO_2 are produced. Coagulase-positive strains produce a variety of toxins and are therefore potentially pathogenic and may cause food poisoning.

Sporolactobacillus sp.

A genus of microaerophilic or anaerobic, nonsporeforming, ordinarily nonmotile bacteria (family Lactobacillaceae) containing gram-positive rods which very from long and slender cells to short coccobacilli; chains are commonly produced, especially in the later part of the logarithmic phase of growth. These organisms possess complex nutritional requirements, generally characteristic for each species; metabolism is fermentative and at least half of the end product is lactic acid. They are found in dairy products and effluents to grain and meat products, water, sewage, beer, wine, fruits and fruit juices, pickled vegetables, and in sour dough and mash, and are part of the normal flora of the mouth, intestinal tract, and vagina of many warm-blooded animals, including humans; rarely are they pathogenic.

Klebssilla pneumoniae

A genus of aerobic, facultatively anaerobic, nonmotile, nonsporeforming bacteria (family Enterobacteriaceae) containing gram-negative, encapsulated rods which occur singly, in pairs, or in short chains. These organisms produce acetylmethylcarbinol and lysine decarboxylase or ornithine decarboxylase. They do not usually liquefy gelatin. Citrate and glucose are ordinarily used as sole carbon sources. These organisms may or may not be pathogenic. They occur in the respiratory, intestinal, and urogenital tracts of man as well as in soil, water, and grain.

Micrococcus luteus

A genus of bacteria (family Micrococcaceae) containing gram-positive, spherical cells that occur in irregular masses, never in packets. Some species are motile or produce motile mutants. These organisms are saprophytic, facultatively parasitic, or parasitic but are not truly pathogenic.

SAMPLING STATION A (Har-ki-Pauri)

(a) Plankton

Quantitative analysis of plankton at this sampling point revealed that the total planktonic concentration was highest 4473 unit/l ±2413.36 in the winter season of 1999-2000 and lowest was 945 unit/l ±615.89 in monsoon season of 1999-2000.

Zooplankton was observed minimum in the monsoon season of 1998-1999 (9 unit/l ±6.34) and maximum in the winter season of 1999-2000 (530 unit/l ±23.67). In 1999-2000 phytoplankton was observed both lower and higher in the season of monsoon and winter respectively (908 unit/l ±497.57 and 3943 unit/l ±724.83) (Table 4B.1 and Figs. 4B.1, 4B.2).

1. Zooplankton

Among zooplankton **Rotifera** was dominating and **Copepoda** was least in both year. At this sampling station **Protozoa** was maximum in winter season of 1999-2000 (142 unit/l ±25.46) and minimum was found Nil in both year's monsoon season. **Rotifera** was found to be highest in the summer season of 1998-1999 (255 unit/l ±8.89) and lowest 2 unit/l ±1.15 in the monsoon season of second year (1999-2000). **Cladocera** was observed in their optimum peak during 1998-1999 winter season (137 unit/l ±10.61) and in lowest peak during monsoon (5 unit/l ±3.54) of the same year (1998-1999). **Copepoda** was maximum 99 unit/l ±3.05 in 1999-2000, winter season and minimum 6 unit/l ±1.50 in summer season of 1999-2000. In total at the sampling station, winter season 1999-2000 showed maximum (530 unit/l ±23.69) and minimum zooplankton in monsoon season of 1998-1999 (29 unit/l ±6.35) (Table 4B.2 and Fig. 4B.3).

The average value revealed the *Philodima* was dominating in the zooplankton, 1998-1999 (45 unit/l ±44.38) followed by *Daphnia* (44 unit/l ±39.40), *Notholca* (43 unit/l ±38.84), *Cyclops* (40 unit/l ±21.39) *Arcella* (36 unit/l ±31.47), *Keratella* (35 unit/l ±46.03), *Amoeba* (33 unit/l ±32.02) and *Bosmina* (23 unit/l ±33.23). In second year 1999-2000 *Amoeba* (50 unit/l ±45.63) and *Keratella* (50 unit/l ±49.03) found in strength followed by *Daphnia* (42 unit/l ±28.21), *Cyclops* (42 unit/l ±49.93), *Bosmina* (37 unit/l ±45.74), *Philodima* (37 unit/l ±46.92), *Arcella* (33 unit/l ±28.57) and *Notholca* (23 unit/l ±28.73) (Table 4B.3 and Fig. 4B.4).

2. Phytoplankton

In phytoplankton **Bacillariophyceae** was dominating and class **Chlorophyceae** was least in both year. At the sampling station **Bacillariophyceae** was observed maximum in winter season (2162 unit/l ±96.41) and minimum in monsoon season (849 unit/l ±40.55) of 1998-1999. **Chlorophyceae** was found to be highest in winter season of 1998-1999 (659 unit/l ±23.33) and lowest (2 unit/l ±0.60) in monsoon season of 1999-2000.

Table 4B.1. Seasonal Variation in Plankton for River Ganga at Sampling Station A (1998-2000)

Plankton	1998 to 1999				1999 to 2000			
	Summer	Monsoon	Winter	Average	Summer	Monsoon	Winter	Average
1. Zooplankton l^{-1}	453	29	413	298.33	378	37	530	315.00
	± 95.93	± 6.34	± 35.98	± 234.11	± 69.91	± 9.98	± 23.67	± 252.47
2. Phytoplankton l^{-1}	2877	1013	3844	2578.00	1504	908	3943	2118.33
	± 472.65	± 444.46	± 784.09	± 1438.99	± 378.15	± 497.57	± 724.83	± 1608.06
Total l^{-1}	3330	1042	4257	2876.33	1882	945	4473	2433.33
	± 1714.03	± 695.79	± 2426.08	± 1654.82	± 796.20	± 615.89	± 2413.36	± 1827.48

± = Standard Deviation.

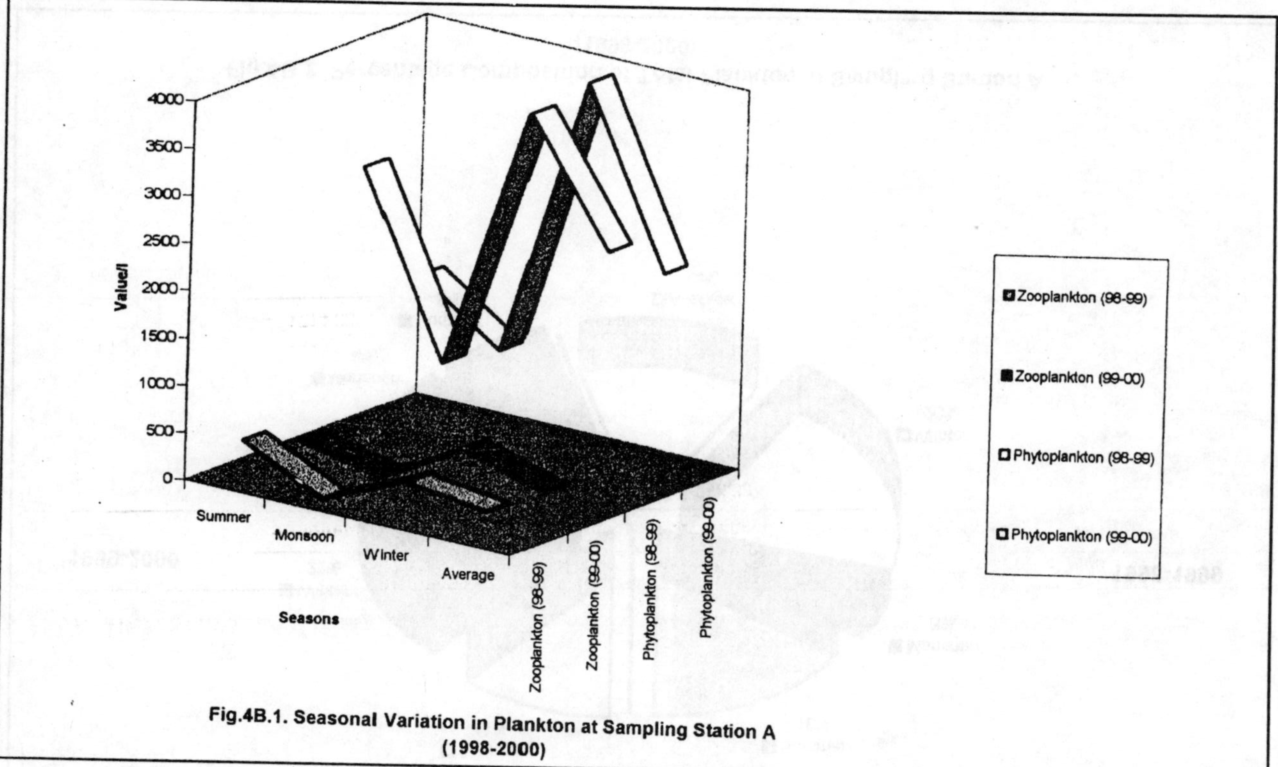

Fig.4B.1. Seasonal Variation in Plankton at Sampling Station A (1998-2000)

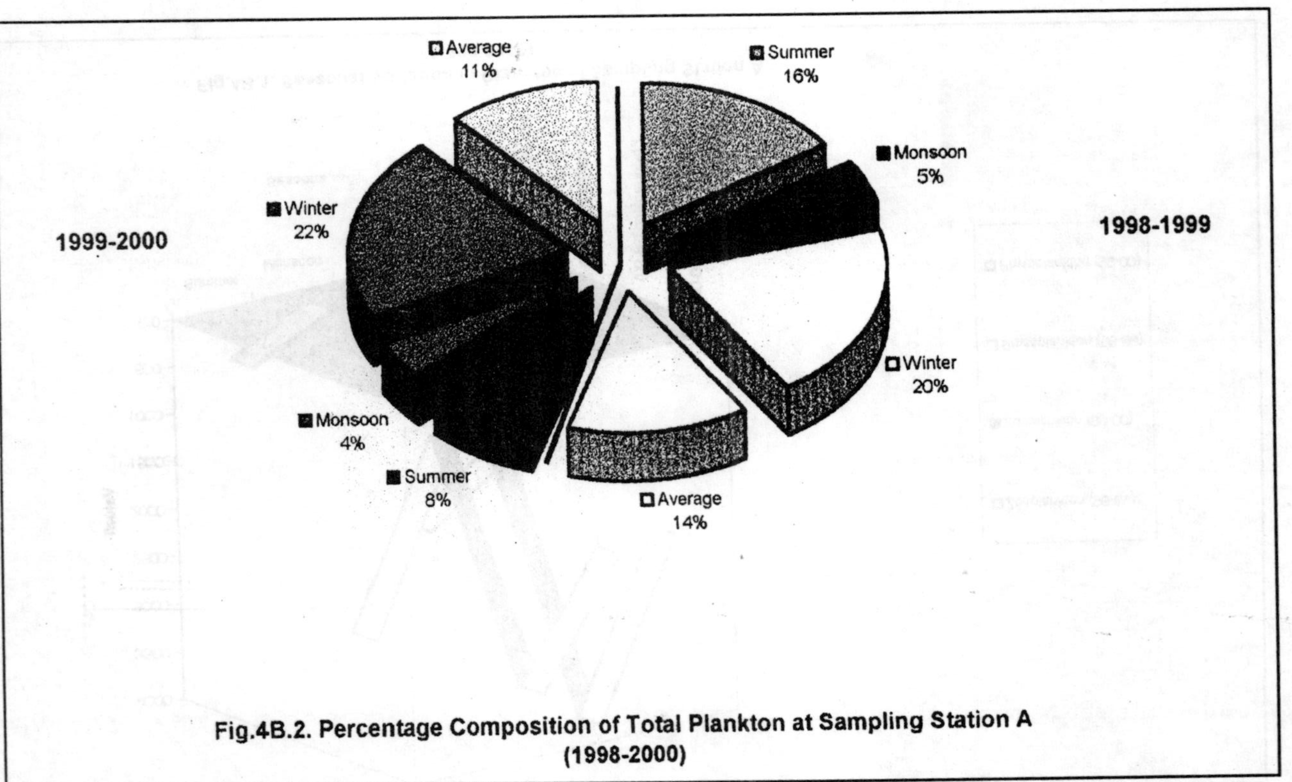

Fig.4B.2. Percentage Composition of Total Plankton at Sampling Station A (1998-2000)

Table 4B.2. Quantitative Analysis for Zooplankton in Different Seasons of River Ganga at Sampling Station A (1998-2000)

	Seasons	Protozoa	Rotifera	Cladocera	Copepoda	Total l^{-1}
1998 to 1999	Summer	89 ± 14.85	255 ± 8.89	58 ± 38.18	51 ± 16.51	453 ± 95.93
	Monsoon	0 ± 0.00	9 ± 2.65	5 ± 3.54	15 ± 9.67	29 ± 6.34
	Winter	118 ± 7.07	105 ± 21.66	137 ± 10.61	53 ± 4.91	413 ± 35.98
	Average l^{-1}	69.00 ± 61.49	123.00 ± 123.98	66.67 ± 66.43	39.67 ± 21.39	298.33 ± 34.92
1999 to 2000	Summer	107 ± 12.02	176 ± 38.07	89 ± 33.23	6 ± 1.50	378 ± 69.91
	Monsoon	0 ± 0.00	2 ± 1.15	14 ± 7.07	21 ± 3.16	37 ± 9.98
	Winter	142 ± 25.46	154 ± 39.00	135 ± 34.41	99 ± 3.05	530 ± 23.67
	Average l^{-1}	83.00 ± 73.98	110.67 ± 94.75	79.33 ± 61.08	42.00 ± 49.93	315.00 ± 28.21

± = Standard Deviation.

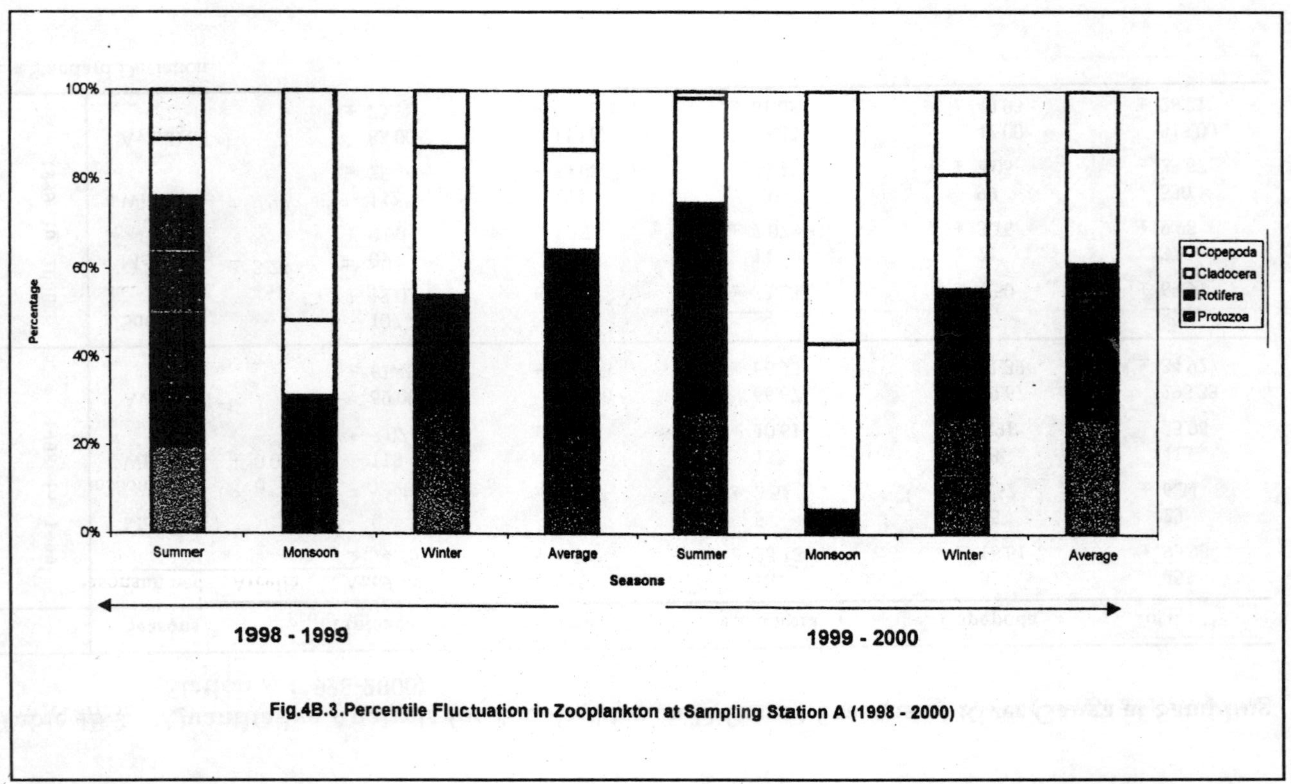

Fig.4B.3.Percentile Fluctuation in Zooplankton at Sampling Station A (1998 - 2000)

Table 4B.3. Quantitative Analysis for Zooplankton in Different Seasons of River Ganga at Sampling Station A (1998-2000)

		Protozoa		Rotifera			Cladocera		Copepoda	
	Seasons	Arcella	Amoeba	Keratella	Philodima	Notholca	Daphnia	Bosmina	Cyclops	Total I $^{-1}$
1998 to 1999	Summer	55 ± 10.26	34 ± 4.88	88 ± 2.09	92 ± 5.27	75 ± 12.56	56 ± 18.21	2 ± 1.00	51 ± 16.51	453 ± 29.53
	Monsoon	0 ± 0.00	0 ± 0.00	5 ± 1.06	4 ± 1.52	0 ± 0.00	0 ± 0.00	5 ± 2.05	15 ± 9.67	29 ± 5.15
	Winter	54 ± 2.36	64 ± 2.05	12 ± 1.26	38 ± 2.51	55 ± 5.08	76 ± 6.06	61 ± 2.16	53 ± 4.91	413 ± 19.31
	Average It $^{-1}$	36 ± 31.47	33 ± 32.02	35 ± 46.03	45 ± 44.38	43 ± 38.84	44 ± 39.40	23 ± 33.23	40 ± 21.39	298 ± 7.40
1999 to 2000	Summer	45 ± 3.26	62 ± 5.02	98 ± 8.16	22 ± 2.05	56 ± 6.27	68 ± 3.34	21 ± 2.06	6 ± 1.50	378 ± 30.08
	Monsoon	0 ± 0.00	0 ± 0.00	0 ± 0.00	0 ± 0.00	2 ± 1.00	12 ± 5.06	2 ± 1.00	21 ± 3.16	37 ± 7.76
	Winter	53 ± 2.36	89 ± 6.02	52 ± 3.16	90 ± 2.19	12 ± 1.05	46 ± 3.05	89 ± 2.44	99 ± 3.05	530 ± 30.25
	Average I $^{-1}$	33 ± 28.57	50 ± 45.63	50 ± 49.03	37 ± 46.92	23 ± 28.73	42 ± 28.21	37 ± 45.74	42 ± 49.93	315 ± 8.92

± = Standard Deviation.

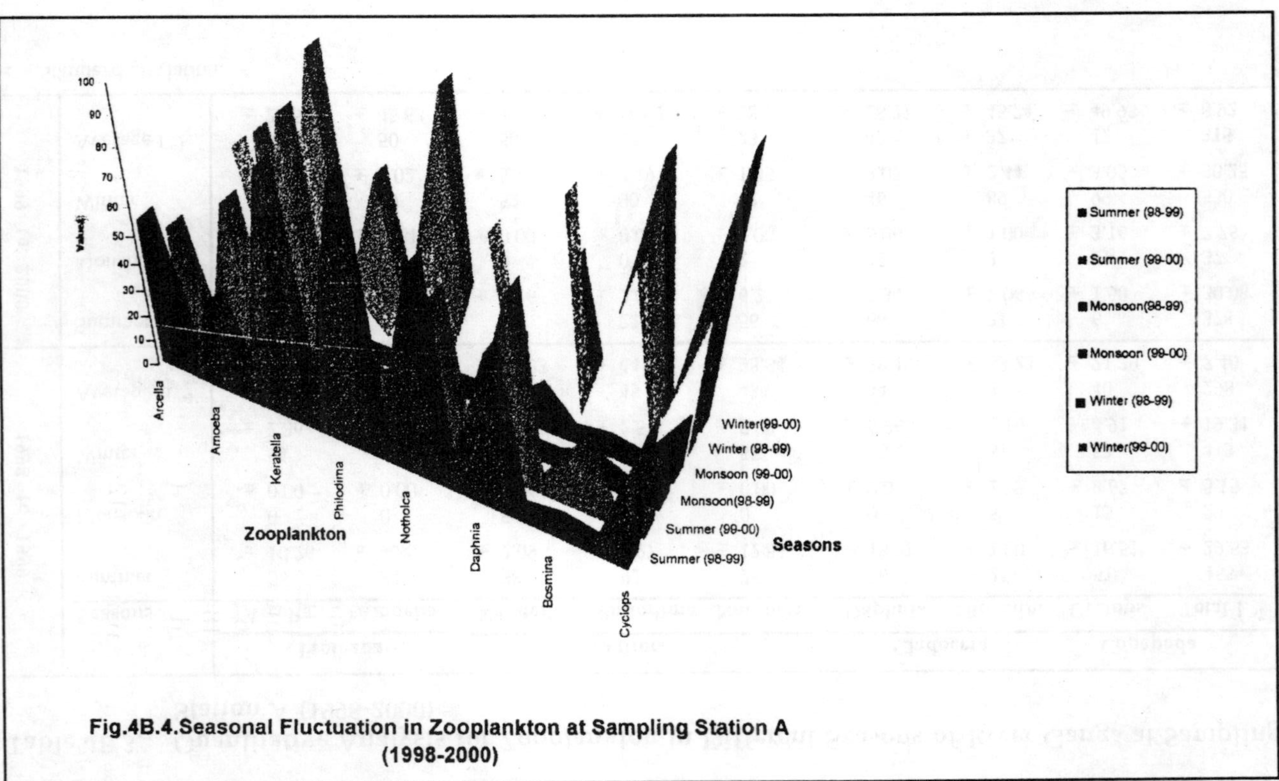

Fig.4B.4.Seasonal Fluctuation in Zooplankton at Sampling Station A (1998-2000)

Myxophyceae showed optimum peak during winter season of 1999-2000 (1576 unit/l ±40.27) and least peak during monsoon season of same year i.e. 1999-2000 (29 unit/l ±7.80) (Table 4B.4 and Figs. 4B.5, 4B.6).

After observing mean values it was noted that *Diatoms* was dominated (231 unit/l ±115.68) among **Bacillariophyceae** in 1998-1999 followed by *Gomphonema,* (214 unit/l ±84.31), *Amphora* (155.67 unit/l ±55.64), *Achnanthes* (151.33 unit/l ±93.43), *Fragilaria* (126.67 unit/l ±84.25), *Cymbella* (93.67 unit/l ±55.59), *Frustulia* (80.33 unit/l ±36.69), *Pinnularia* (72.33 unit/l ±35.64), *Tabellaria* (71.33 unit/l ±35.50), *Surirella* (60.33 unit/l ±55.10), *Navicula* (48 unit/l ±26.91), *Synedra* (39 unit/l ±35.37), *Stauroneis* (19.67 unit/l ±19.66), *Cocconeis* (19.33 unit/l ±31.77), *Cyclotella* (17.67 unit/l ±26.39), *Phacus* (17 unit/l ±12.53) and *Nitzschia* (9.33 unit/l ±9.02). In second year 1999-2000, again *Diatoms* (172 unit/l ±115.64) were recorded as a dominating followed by *Amphora* (159 unit/l ±87.48), *Fragilaria* (150.30 unit/l ±121.50), *Gomphonema* (137.33 unit/l ±56.08), *Achnanthes* (101.67 unit/l ±48.44), *Frustulia* (72.33 unit/l ±49.37), *Surirella* (66 unit/l ±44.54), *Tabellaria* (63.33 unit/l ±38.94), *Synedra* (49.67 unit/l ±54.99), *Navicula* (46.33 unit/l ±19.60), *Stauroneis* (44 unit/l ±22.54), *Cocconeis* (41 unit/l ±58.55), *Cymbella* (39.33 unit/l ±57.87), *Pinnularia* (24 unit/l ±38.16), *Cyclotella* (22 unit/l ±38.11), *Phacus* (20.33 unit/l ±23.46) and *Nitzschia* (16.33 unit/l ±15.04) (Table 4B. 5 and Fig. 4B.7).

A great fluctuation among **Bacillariophyceae** was observed in seasonal study at this sampling point. In 1998-1999 summer season, *Diatoms* was recorded highest (250 unit/l ±14.91) and *Cocconeis* were found to be Nil; in monsoon season *Gomphonema* was maximum (118 unit/l ±10.12) and *Cyclotella* was observed to be Nil. In winter again optimum concentration was observed for *Diatoms* (336 unit/l ±9.22) and least concentration was noted for *Nitzschia* (18 unit/l ±0.08). In 1999-2000 summer, *Diatoms* were recorded highest (158 unit/l ±5.29) and *Cyclotella, Pinnularia, Phacus* were recorded Nil; in monsoon season *Synedra* was noted maximum (112 unit/l ±10.62) while *Cocconeis* and *Cyclotella* were recorded Nil; in winter season *Diatoms* was recorded in optimum concentration (294 unit/l ±2.09) and *Nitzschia* was observed least concentrated (2 unit/l ±0.05).

Table 4B.4. Comparative Analysis for Seasonal Variation Among Phytoplankton of River Ganga at Sampling Station A (1998-2000)

	Seasons	Bacillariophyceae l^{-1}	Chlorophyceae l^{-1}	Myxophyceae l^{-1}	Total l^{-1}
1998 to 1999	Summer	1269 ± 83.19	415 ± 16.79	1193 ± 60.39	2877 ± 472.65
	Monsoon	849 ± 40.55	44 ± 5.46	120 ± 23.89	1013 ± 444.46
	Winter	2162 ± 96.41	659 ± 23.33	1023 ± 156.05	3844 ± 784.09
	Average	1426.67 ± 670.55	372.67 ± 309.68	778.67 ± 576.72	2578 ± 531.61
1999 to 2000	Summer	926 ± 58.45	201 ± 13.52	377 ± 40.02	1504 ± 378.15
	Monsoon	877 ± 43.76	2 ± 0.60	29 ± 7.80	908 ± 497.57
	Winter	1872 ± 90.62	495 ± 20.23	1576 ± 40.27	3943 ± 724.83
	Average	1225.00 ± 560.85	232.67 ± 248.02	660.67 ± 811.57	2118 ± 497.73

± = Standard Deviation.

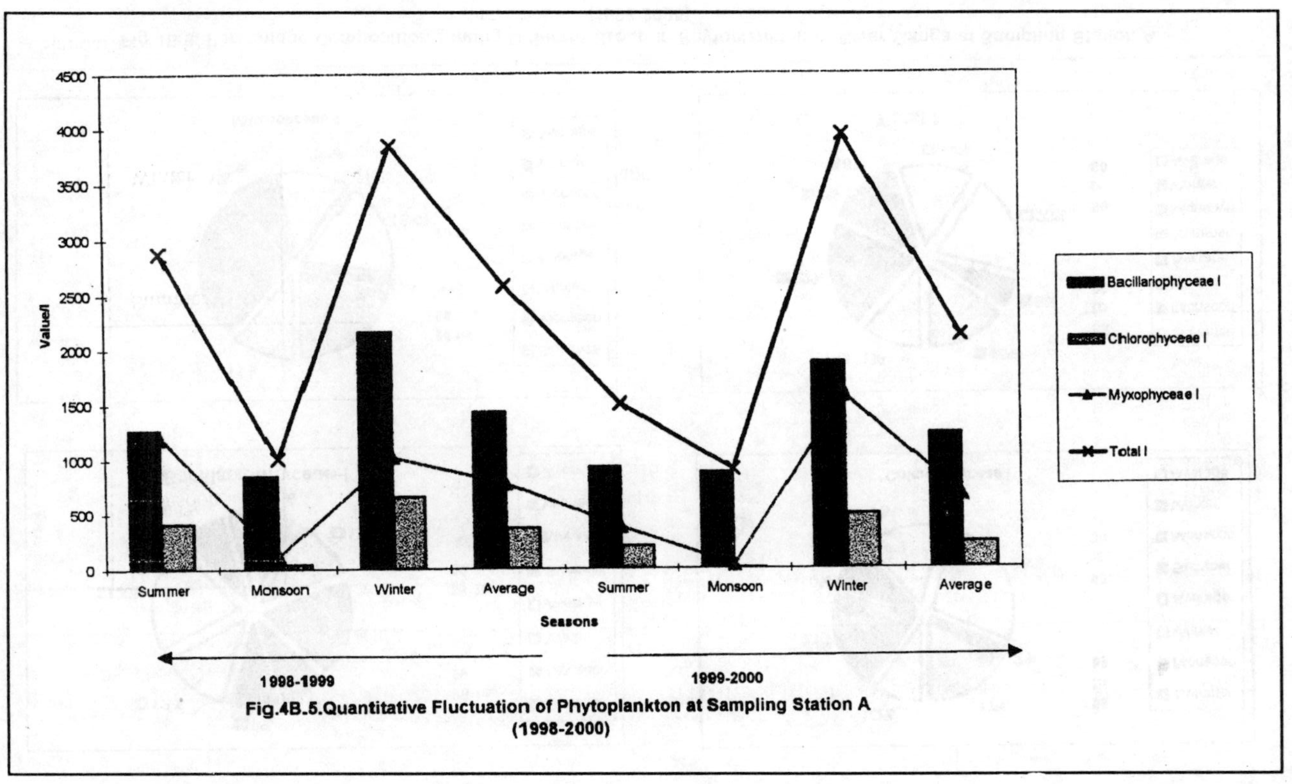

Fig.4B.5.Quantitative Fluctuation of Phytoplankton at Sampling Station A (1998-2000)

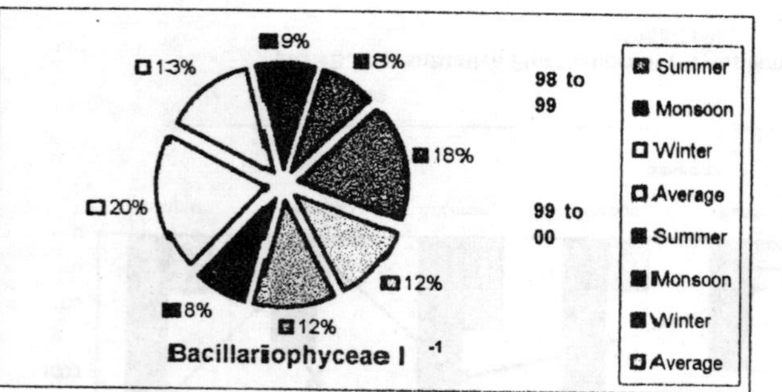

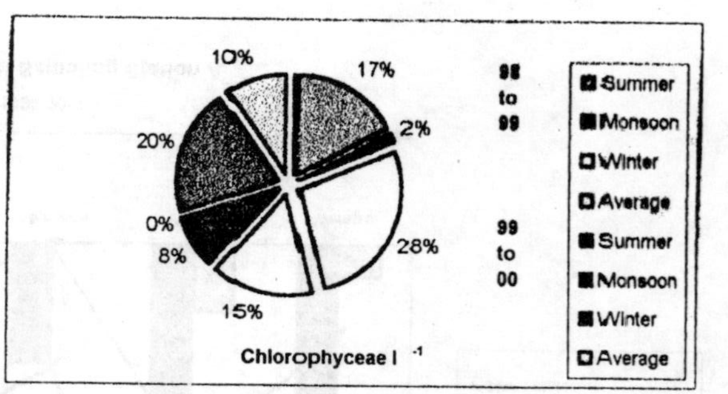

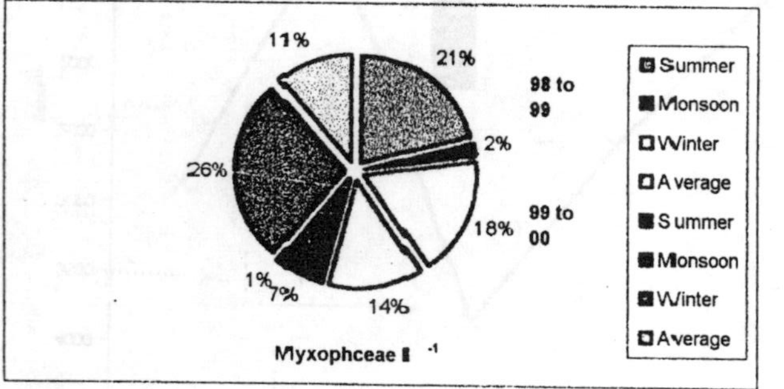

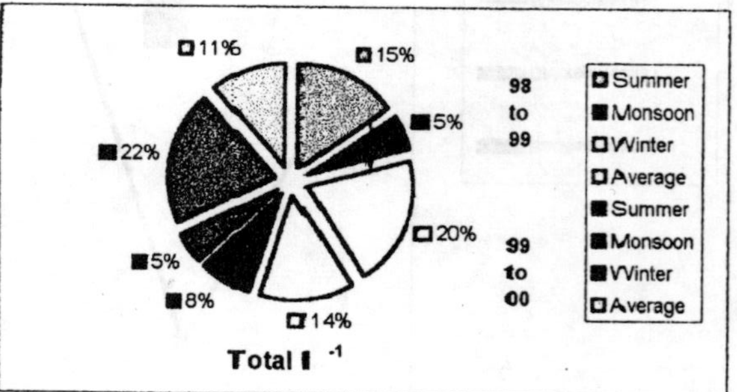

Fig 4B.6. Percentage Composition Among Different Group in Phytoplankton of River Ganga at Sampling Station A (1998-2000)

Table 4B.5. Quantitative Analysis for Seasonal Variation Among *Bacillariophyceae* of River Ganga at Sampling Station A (1998-2000)

	Seasons	Achnanthes	Amphora	Cocconeis	Cymbella	Cyclotella	Diatoms	Fragilaria	Frustulia	Gomphonena
1998 to 1999	Summer	112	180	0	90	5	250	110	76	248
		± 12.11	± 10.01	± 0.00	± 2.22	± 0.01	± 14.91	± 10.25	± 1.22	± 14.22
	Monsoon	84	92	2	40	0	107	52	46	118
		± 3.59	± 2.09	± 0.01	± 3.10	± 0.00	± 10.11	± 6.46	± 0.05	± 10.12
	Winter	258	195	56	151	48	336	218	119	276
		± 10.46	± 5.91	± 1.22	± 2.49	± 0.09	± 9.22	± 11.91	± 5.59	± 11.59
	Average l^{-1}	151.33	155.67	19.33	93.67	17.67	231.00	126.67	80.33	214.00
		± 93.43	± 55.64	± 31.77	± 55.59	± 26.39	± 115.68	± 84.25	± 36.69	± 84.31
1999 to 2000	Summer	86	107	15	10	0	158	69	59	202
		± 2.19	± 6.49	± 0.05	± 0.08	± 0.00	± 5.29	± 2.73	± 1.05	± 10.38
	Monsoon	63	110	0	2	0	64	92	31	108
		± 2.55	± 5.22	± 0.00	± 0.01	± 0.00	± 1.61	± 2.34	± 2.08	± 9.56
	Winter	156	260	108	106	66	294	290	127	102
		± 5.49	± 11.06	± 10.04	± 8.22	± 2.37	± 2.09	± 14.22	± 12.61	± 2.06
	Average l^{-1}	101.67	159.00	41.00	39.33	22.00	172.00	150.33	72.33	137.33
		± 48.44	± 87.48	± 58.51	± 57.87	± 38.11	± 115.64	± 121.50	± 49.37	± 56.08

Table 4B.5. (Contd.)

	Seasons	Navicula	Nitzschia	Pinnularia	Phacus	Stauroneis	Surirella	Synedra	Tabellaria	Total I $^{-1}$
1998 to 1999	Summer	70 ± 1.09	0 ± 0.00	62 ± 1.06	18 ± 3.44	12 ± 0.05	0 ± 0.00	0 ± 0.00	36 ± 0.05	1269 ± 83.19
	Monsoon	18 ± 0.02	10 ± 0.02	43 ± 2.24	4 ± 0.05	5 ± 0.01	73 ± 2.33	48 ± 2.01	107 ± 9.22	849 ± 40.55
	Winter	56 ± 1.06	18 ± 0.08	112 ± 14.12	29 ± 1.55	42 ± 0.09	108 ± 14.66	69 ± 2.44	71 ± 2.46	2162 ± 92.41
	Average I $^{-1}$	48.00 ± 26.91	9.33 ± 9.02	72.33 ± 35.64	17.00 ± 12.53	19.67 ± 19.66	60.33 ± 55.10	39.00 ± 35.37	71.33 ± 35.50	1426.67 ± 69.40
1999 to 2000	Summer	44 ± 5.16	15 ± 0.05	0 ± 0.00	0 ± 0.00	56 ± 0.08	18 ± 1.08	8 ± 0.00	79 ± 1.05	926 ± 58.45
	Monsoon	28 ± 1.24	32 ± 1.01	4 ± 0.01	15 ± 2.41	18 ± 0.02	106 ± 5.42	112 ± 10.62	92 ± 1.19	877 ± 43.76
	Winter	67 ± 3.33	2 ± 0.05	68 ± 0.09	46 ± 3.11	58 ± 2.42	74 ± 6.66	29 ± 1.06	19 ± 0.03	1872 ± 90.62
	Average I $^{-1}$	46.33 ± 19.60	16.33 ± 15.04	24.00 ± 38.16	20.33 ± 23.46	44.00 ± 22.54	66.00 ± 44.54	49.67 ± 54.99	63.33 ± 38.94	1225.00 ± 52.15

± = Standard Deviation.

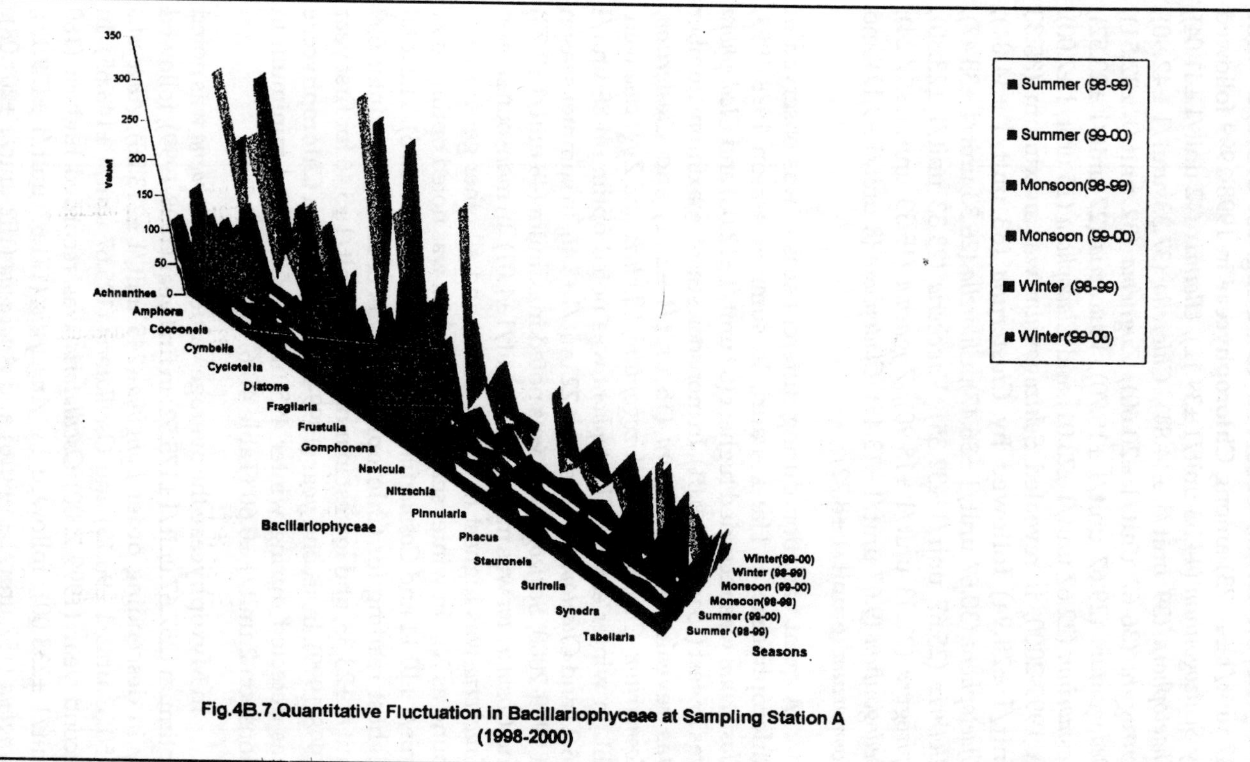

Fig.4B.7.Quantitative Fluctuation in Bacillariophyceae at Sampling Station A (1998-2000)

The *Closterium* observed to be highest average value (47 unit/l ±40.73) among **Chlorophyceae** in 1998-1999 followed by *Schizogonium* (44.33 unit/l ±39.12), *Ulothrix* (42 unit/l ±41.04), *Chaetophora* (39 unit/l ±44.84), *Chlorella* (37.33 unit/l ±42.90), *Spirogyra* (36.67 unit/l ±20.60), *Zygnema* (32 unit/l ±32.51), *Oedogonium* (29.67 unit/l ±15.70), *Vaucheria* (27 unit/l ±23.52), *Cosmarium* (22.67 unit/l ±23.01) and *Cladophora* (15 unit/l ±7.00). In 1999-2000, it revealed *Schizogonium* was maximum (33.33 unit/l ±28.94) followed by *Closterium* (33 unit/l ±34.04), *Chaetophora* (30.67 unit/l ±36.47), *Chlorella* (26.33 unit/l ±33.47), *Ulothrix* (25.67 unit/l ±32.35), *Vaucheria* (23.33 unit/l ±23.80), *Spirogyra* (21.33 unit/l ±18.90), *Zygnema* (15.33 unit/l ±17.24), *Oedogonium* (9.67 unit/l ±13.43), *Cladophora* (8 unit/l ±9.17) and *Cosmarium* (6 unit/l ±8.72).

A great variation during different season was observed in **Chlorophyceae** at Har-ki-Pauri. In summer season 1998-1999, *Closterium* was recorded highest (69 unit/l ±12.08) and *Cladophora* was lowest (15 unit/l ±0.01). In monsoon season maximum number was revealed for *Spirogyra* (15 unit/l ±2.11) and *Closterium, Cosmarium, Chaetophora, Schizogonium, Ulothrix* and *Zygnema* were Nil. In winter season *Chaetophora* was noted optimum 88 unit/l ±6.23 and *Chaetophora* was least 22 unit/l ±8.46. In summer season of 1999-2000, *Schizogonium* was noted maximum (48 unit/l ±2.73) and *Cosmarium* was minimum (2 unit/l ±1.01). In monsoon season *Vaucheria* was highest (2 unit/l ±0.50) and all other genera were found as Nil. In winter season *Chaetophora* was noted optimum 71 unit/l ±10.11 and *Cosmarium* was least (16 unit/l ±5.76). The total highest reading for **Chlorophyceae** was observed in winter 659 unit/l ±23.33, and lowest in monsoon 44 unit/l ±5.46 for first year (1998-1999). In second year (1999-2000) maximum **Chlorophyceae** was detected during winter 495 unit/l ±20.23 and minimum in monsoon 2 unit/l ±0.60 (Table 4B.6, Fig. 4B.8).

In **Myxophyceae** the average value was *Anabaena* was noted optimum (257.67 unit/l ±175.52) in first year (1998-1999), followed by in descending order *Lyngbya* (245 unit/l ±213.26), *Anacystis* (150.33 unit/l ±96.13) and *Oscillatoria* (125.67 unit/l ±143.65). In second year (1999-2000) *Oscillatoria* was recorded highest (180 unit/l ±238.00), followed by *Anacystis* (171.67 unit/l ±179.15), *Lyngbya* (157 unit/l ±199.16) and *Anabaena* (152 unit/l ±201.08).

Table 4B.6. Quantitative Analysis for Seasonal Variation Among *Chlorophyceae* of River Ganga at Sampling Station A (1998-2000)

	Seasons	Closterium	Cosmarium	Chaetophora	Chlorella	Cladophora	Oedogonium
1998 to 1999	Summer	69 ± 12.08	22 ± 2.06	29 ± 1.05	21 ± 2.22	15 ± 0.01	35 ± 5.52
	Monsoon	0 ± 0.00	0 ± 0.00	0 ± 0.00	5 ± 1.06	8 ± 2.07	12 ± 3.44
	Winter	72 ± 10.46	46 ± 5.91	88 ± 6.23	86 ± 7.11	22 ± 8.46	42 ± 3.22
	Average I^{-1}	47.00 ± 40.73	22.67 ± 23.01	39.00 ± 44.84	37.33 ± 42.90	15.00 ± 7.00	29.67 ± 15.70
1999 to 2000	Summer	31 ± 2.64	2 ± 1.01	21 ± 4.06	15 ± 2.56	6 ± 1.05	4 ± 0.51
	Monsoon	0 ± 0.00	0 ± 0.00	0 ± 0.00	0 ± 0.00	0 ± 0.00	0 ± 0.00
	Winter	68 ± 4.38	16 ± 5.76	71 ± 10.11	64 ± 8.22	18 ± 2.05	25 ± 1.06
	Average I^{-1}	33.00 ± 34.04	6.00 ± 8.72	30.67 ± 36.47	26.33 ± 33.47	8.00 ± 9.17	9.67 ± 13.43

Table 4B.6. (Contd.)

	Seasons	Schizogonium	Spirogyra	Ulothrix	Vaucheria	Zygnema	Total I^{-1}
1998 to 1999	Summer	59 ± 10.25	39 ± 1.22	44 ± 10.21	51 ± 4.11	31 ± 2.04	415 ± 16.79
	Monsoon	0 ± 0.00	15 ± 2.11	0 ± 0.00	4 ± 1.25	0 ± 0.00	44 ± 5.46
	Winter	74 ± 4.16	56 ± 3.48	82 ± 6.27	26 ± 2.44	65 ± 5.21	659 ± 23.33
	Average I^{-1}	44.33 ± 39.12	36.67 ± 20.60	42.00 ± 41.04	27.00 ± 23.52	32.00 ± 32.51	372.67 ± 9.72
1999 to 2000	Summer	48 ± 2.73	28 ± 1.28	15 ± 2.34	19 ± 5.16	12 ± 1.22	201 ± 13.52
	Monsoon	0 ± 0.00	0 ± 0.00	0 ± 0.00	2 ± 0.50	0 ± 0.00	2 ± 0.60
	Winter	52 ± 3.49	36 ± 1.08	62 ± 3.19	49 ± 2.01	34 ± 0.52	495 ± 20.23
	Average I^{-1}	33.33 ± 28.94	21.33 ± 18.90	25.67 ± 32.35	23.33 ± 23.80	15.33 ± 17.24	232.67 ± 10.00

± = Standard Deviation.

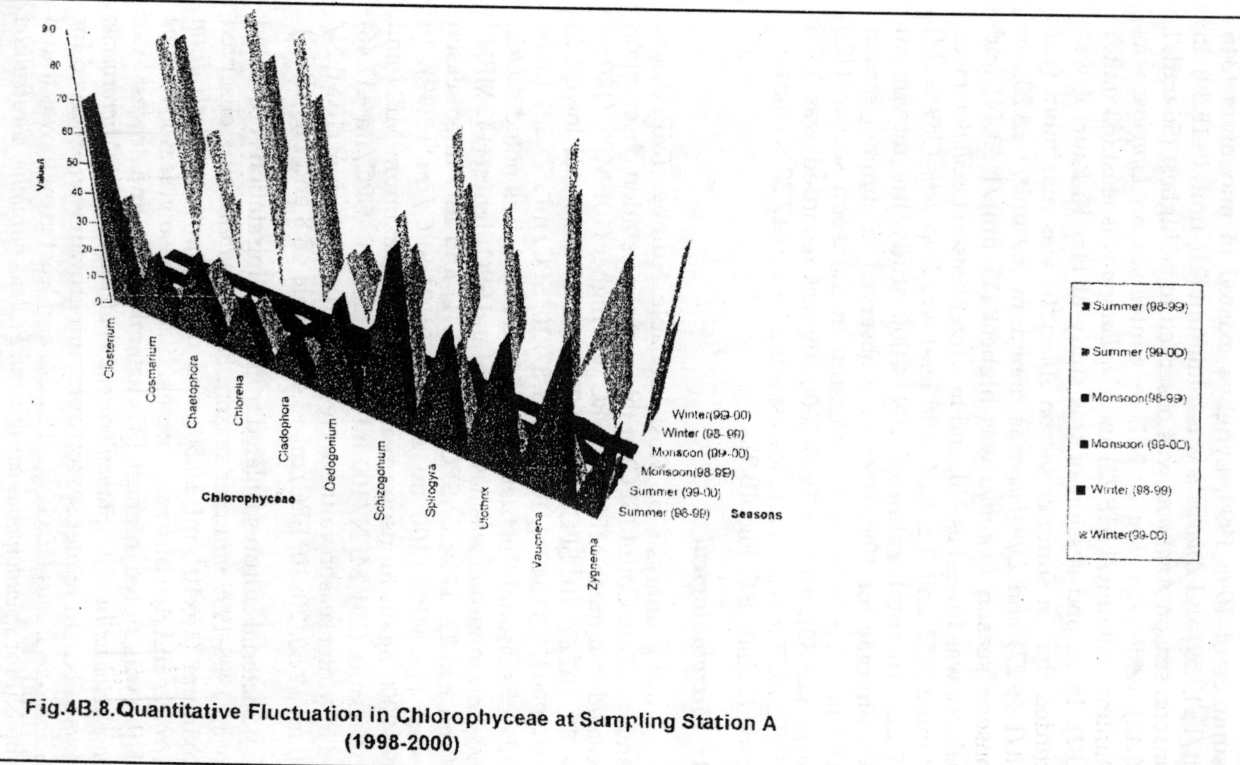

Fig.4B.8. Quantitative Fluctuation in Chlorophyceae at Sampling Station A (1998-2000)

For seasonal fluctuation of **Myxophyceae** it is revealed that is summer of 1998-1999, Lyngbya existed in maximum (346 unit/l ±17.59) and *Anacystis* was minimum (216 unit/l ±15.39). In monsoon season *Anabaena* was observed to be highest (56 unit/l ±16.42), while *Lyngbya* as Nil. In winter season *Lyngbya* was optimum (389 unit/l ±18.22) and *Oscillatoria* was least (63 unit/l ±5.82). In second session of this study at Har-Ki-Pauri it was recorded that in summer season *Anacystis* was maximum (154 unit/l ±6.27) and *Lyngbya* was minimum (69 unit/l ±8.20), in monsoon season *Lyngbya* was highest (17 unit/l ±3.44) and *Anabaena* was lowest as Nil, and in winter season *Oscillatoria* was optimum (452 unit/l ±20.34) and least was *Anacystis* (359 unit/l ±17.22). In total value of 1998-2000 study the maximum **Myxophyceae** for 1998-1999 were observed in summer season (1193 unit/l ±60.39) and minimum in monsoon season (120 unit/l ±23.89), and in 1999-2000 highest recorded was 1576 unit/l ±40.27 in winter and lowest was 29 unit/l ±7.80 in monsoon season (Table 4B.7, Fig. 4B.9).

(b) Microbiological

During analysis of Most Probable Number (MPN) and Standard Plate Count (SPC) at this sampling station observation revealed that mean MPN and SP.C was higher (101 MPN/100 ml ±90.50, 102.67×10^3 SPC/ml ±94.71) in 1999-2000 and lowest in 1998-1999 (73.33 MPN/100ml ±65, 87×10^3 SPC/ml ± 77.13). In the study of seasonal fluctuation at Har-ki-Pauri, it is noted that in 1998-1999 monsoon was having MPN and SPC highest (142 MPN/100 ml ±15.22, 174×10^3 SPC/ml ± 13.41) and in winter season lowest (12 MPN/100 ml ±5.24, 27×10^3 SPC/ml ±2.99). In 1999-2000 again monsoon season showed maximum MPN and SPC results (191 MPN/100 ml ±14.09, 206×10^3 SPC/ml ±12.46) and in winter season values were recorded as minimum (10 MPN/100 ml ±2.50, 20×10^3 SPC/ml ±3.40) (Table 4B.8 and Fig. 4B.10).

Bacterial cultures isolated from sampling station A revealed that in 1998-1999 summer season, *Bacillus subtilis* was observed maximum (15×10^3/ml ±0.50) and minimum was *Clostridium perfringes* and *Aerobacter aerogens* as Nil. In monsoon season *Bacillus subtilis* was noted highest (35×10^3/ml ±3.44) and lowest was *Sporolactobacillus sp., Staphylococcus epidermidis* and *Micrococcus luteus* as Nil. In winter season optimum growth was observed for *Micrococcus luteus* (10×10^3/ml ±2.24) and least growth was found to be Nil for *Pseudomonas aeruginosa, Proteus mirabilis, Clostridium*

Table 4B.7. Quantitative Analysis For Seasonal Variation Among *Myxophyceae* of River Ganga at Sampling Station A (1998-2000)

	Seasons	Anacystis	Anabaena	Lyngbya	Oscillatoria	Total I^{-1}
1998 to 1999	Summer	216 ± 15.39	341 ± 18.21	346 ± 17.59	290 ± 16.51	1193 ± 60.39
	Monsoon	40 ± 10.22	56 ± 16.42	0 ± 0.00	24 ± 9.67	120 ± 23.89
	Winter	195 ± 10.46	376 ± 15.91	389 ± 18.22	63 ± 5.82	1023 ± 156.05
	Average I^{-1}	150.33 ± 96.13	257.67 ± 175.52	245.00 ± 213.26	125.67 ± 143.65	778.67 ± 66.41
1999 to 2000	Summer	154 ± 6.27	76 ± 4.31	69 ± 8.20	78 ± 6.11	377 ± 40.02
	Monsoon	2 ± 1.00	0 ± 0.00	17 ± 3.44	10 ± 3.06	29 ± 7.80
	Winter	359 ± 17.22	380 ± 18.53	385 ± 18.82	452 ± 20.34	1576 ± 40.27
	Average I^{-1}	171.67 ± 179.15	152.00 ± 201.08	157.00 ± 199.16	180.00 ± 238.00	660.67 ± 12.94

± = Standard Deviation.

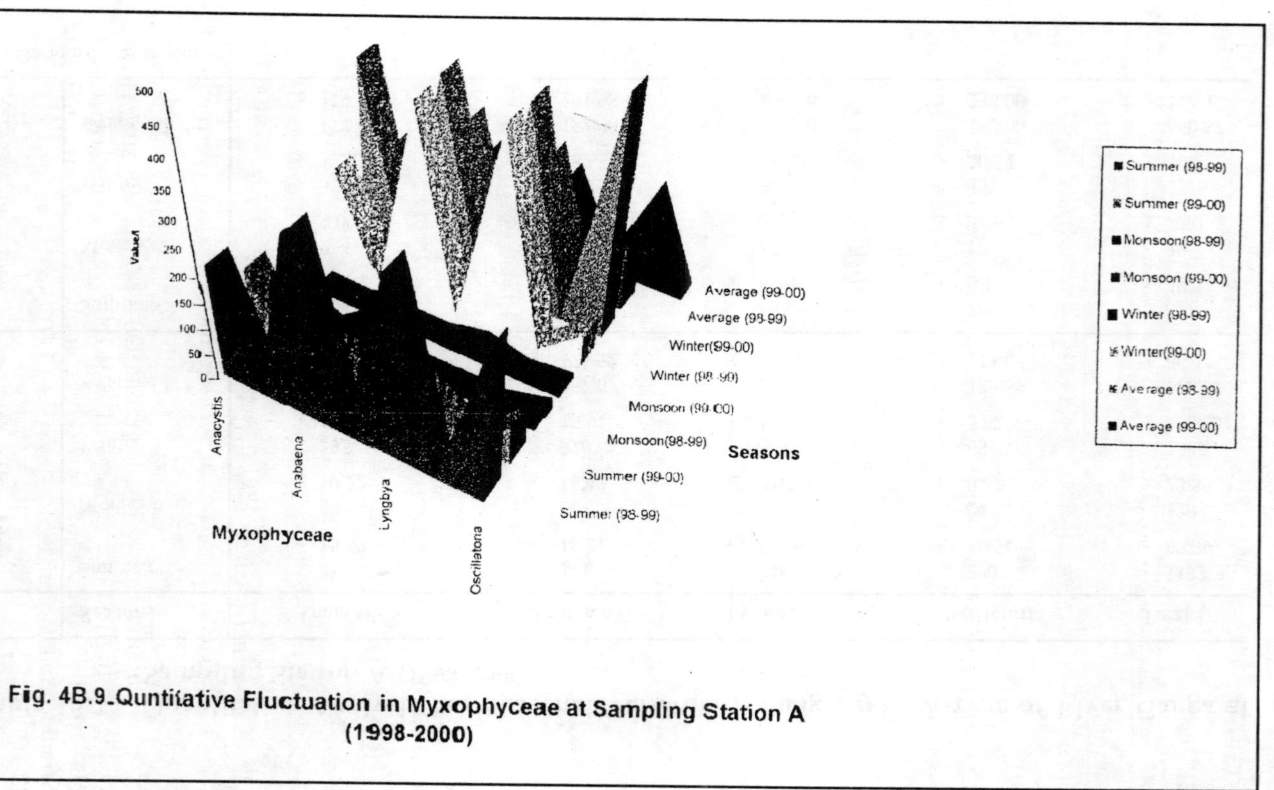

Fig. 4B.9. Quntitative Fluctuation in Myxophyceae at Sampling Station A (1998-2000)

Table 4B.8. Quntitative Analysis for Seasonal Variation in Bateriological Parameters of River Ganga at Sampling Station A (1998-2000)

Bacteriology		MPN/100 ml	SPC/ml (×10³)
1998 To 1999	Summer	78 ± 12.05	60 ± 4.39
	Monsoon	142 ± 15.22	174 ± 13.41
	Winter	12 ± 5.24	27 ± 2.99
	Average	77.33 ± 65.00	87.00 ± 77.13
1999 To 2000	Summer	102 ± 11.53	82 ± 7.74
	Monsoon	191 ± 14.09	206 ± 12.46
	Winter	10 ± 2.50	20 ± 2.57
	Average	101.00 ± 90.50	102.67 ± 94.71

± = Standard Deviation.

perfringes, Aerobacter aerogens, Salmonella typhii, Klebssilla pneumoniae and *Staphylococcus epidermidis*. In 1999-2000 summer season's observation revealed that growth of *Bacillus subtilis* was optimum ($25x10^3$/ml ±2.56) and *Proteus mirabilis, Clostridium perfringes* and *Aerobacter aerogens* was absent. In monsoon season growth of *Streptococcus facecalis* was recorded as highest (38×10^3/ml ±2.46) and *Klebssilla pneumoniae* was lowest as Nil. In winter season *Bacillus subtilis* showed optimum growth (8×10^2/ml ±0.50) and *Proteus mirabilis, Clostridium perfringes, Streptococcus facecalis, Aerobacter aerogens, Salmonella typhii, Klebssilla pneumoniae, Staphylococcus epidermidis,* and *Micrococcus luteus* were recorded as Nil.

On an average calculation for 1998-1999 *Bacillus subtilis* was observed highest (18×10^3/ml ±15.28) followed by *Escherichia coli* (14×10^3/ml ±13.87), *Staphylococcus anreus* (14×10^3/ml ±17.32), *Streptococcus facecalis* (10×10^3/ml ±13.00) *Pseudomonas aeruginosa* (7 10^3/ml ±11.02), *Micrococcus luteus* (6×10^3/ml ±2.31), *Salmonella typhii* (6 $x10^3$/ml ±6.03), *Sporolactobacillus sp.* (3×10^3/ml ±3.51),

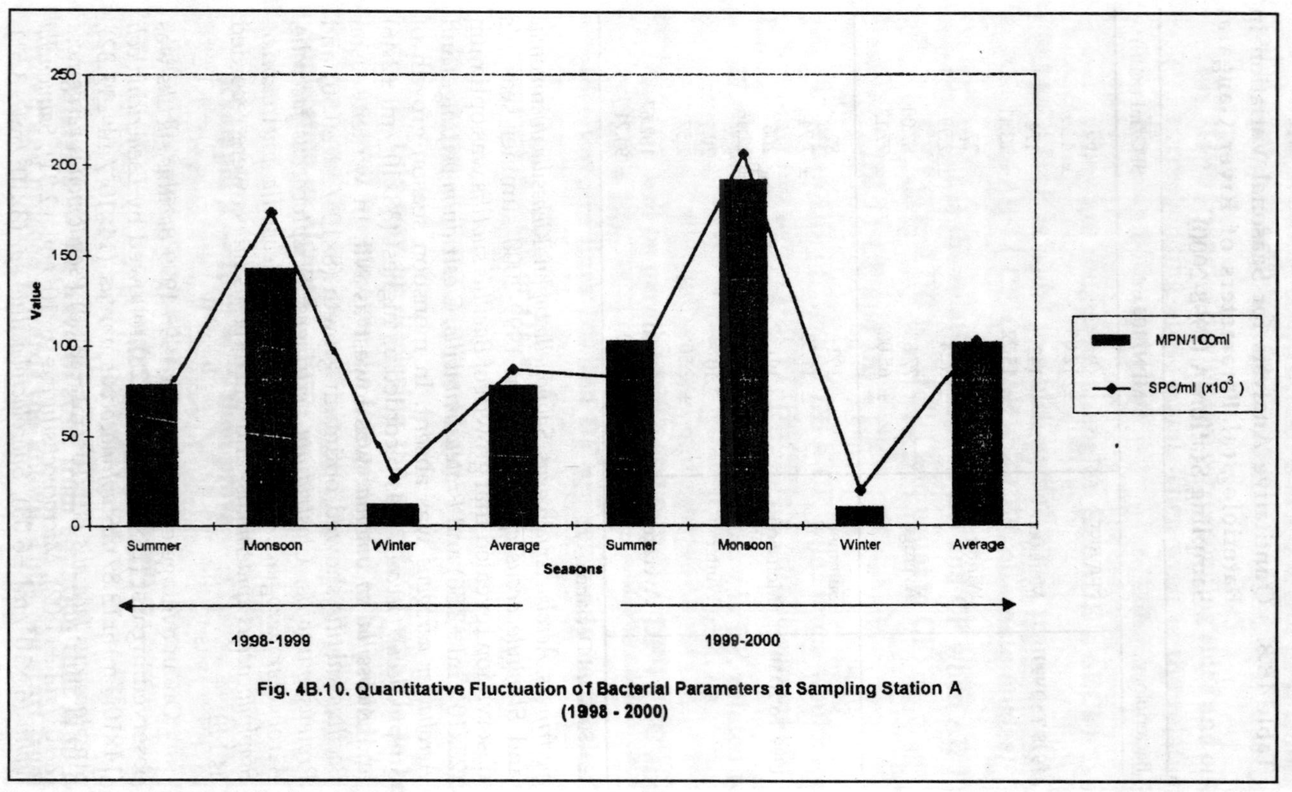

Fig. 4B.10. Quantitative Fluctuation of Bacterial Parameters at Sampling Station A (1998 - 2000)

Klebssilla pneumoniae (2×10^3/ml ±2.52), *Proteus mirabilis* (2×10^3 /ml ±2.65), *Aerobacter aerogens* (2×10^3/ml ±3.46), *Clostridium perfringes* (1×10^3 /ml ±1.73) and *Staphylococcus epidermidis* (1×10^3 /ml ±2.31). In 1999-2000 *Bacillus subtilis* was noted maximum (15 $\times10^3$/ml ±8.89) followed by *Escherichia coli* (15×10^3/ml ±11.79), *Pseudomonas aeruginosa* (14×10^3/ml ±19.16), *Streptococcus facecalis* (13×10^3/ml ±21.39), *Salmonella typhii* (10×10^3/ml ±13.05), *Staphylococcus epidermidis* (8×10^3/ml ±7.51), *Staphylococcus aureus* (7 $\times10^3$/ml ±9.24), *Aerobacter aerogens* (6×10^3/ml ±10.39), *Sporolactobacillus sp.* (5×10^3/ml ±4.16), *Micrococcus luteus* (5×10^3/ml ±7.51), *Clostridium perfringes* (2×10^3/ml ±2.89) and *Klebssilla pneumoniae* was found to be Nil in this session (Table 4B.9 and Fig. 4B.11).

SAMPLING STATION B (Mayapuri Ghat)

(a) Plankton

Quantitative analysis of plankton at Mayapuri Ghat showed that the total planktonic concentration was maximum 4729 unit/l ±2383.66 in the winter season of 1998-1999 and minimum was 493 unit/l ±348.60 in monsoon season of the same 1999-2000.

Zooplankton was noted lowest as Nil in the monsoon season of 1999-2000 and highest in the winter season of 1999-2000 (679 unit/l ±79.58) Phytoplankton was observed minimum in monsson season of 1999-2000 (493 unit/l ±86.07) and maximum in 1998-1999 for the season of winter (4050 unit/l ±1504.99) (Table 4B.10 and Figs. 4B.12, 4B.13).

1. Zooplankton

Among zooplankton phylum **Rotifera** was dominating and phylum **Copepoda** was least in both year. At this spot **Protozoa** was maximum in winter season of 1998-1999 (187 unit/l ±16.26) and minimum was found Nil in both year's monsoon season. **Rotifera** was found to be highest in the summer season of 1998-1999 (274 unit/l ±3.06) and lowest as Nil in the monsoon season of second year (1999-2000) **Cladocera** was observed in their optimum peak during 1998-1999 winter season (124 unit/l ±18.38) and in lowest peak during monsoon season as Nil in 1999-2000. **Copepoda** was maximum (94 unit/l ±7.51) in 1998-1999, winter season and minimum was found Nil in monsoon

Table 4B.9. Isolation of Bacterial Culture from Sampling Station A (1998-2000)

Bacteria	1998 to 1999				1999 to 2000			
	Summer	Monsoon	Winter	Average	Summer	Monsoon	Winter	Average
Escherichia coli*	10 ± 1.50	29 ± 3.20	2 ± 0.50	14 ± 13.87	12 ± 1.50	28 ± 2.22	5 ± 1.00	15 ± 11.79
Pseudomonas aeruginosa*	2 ± 0.50	20 ± 2.51	0 ± 0.00	7 ± 11.02	5 ± 2.10	36 ± 1.52	1 ± 0.50	14 ± 19.16
Bacillus subtilis**	15 ± 0.50	35 ± 3.44	5 ± 1.20	18 ± 12.28	25 ± 2.56	12 ± 1.11	8 ± 0.50	15 ± 8.89
Proteus mirabilis**	1 ± 0.00	5 ± 0.56	0 ± 0.00	2 ± 2.65	0 ± 0.00	6 ± 0.50	0 ± 0.00	2 ± 3.46
Clostridium perfringes*	0 ± 0.00	3 ± 1.00	0 ± 0.00	1 ± 1.73	0 ± 0.00	5 ± 0.50	0 ± 0.00	2 ± 2.89
Streptococcus facecalis*	2 ± 1.50	25 ± 2.11	3 ± 0.51	10 ± 13.00	2 ± 0.50	38 ± 2.46	0 ± 0.00	13 ± 21.39
Aerobacter aerogens*	0 ± 0.00	6 ± 0.53	0 ± 0.00	2 ± 3.46	0 ± 0.00	18 ± 1.50	0 ± 0.00	6 ± 10.39
Salmonella typhii*	5 ± 0.52	12 ± 1.20	0 ± 0.00	6 ± 6.03	6 ± 1.00	25 ± 1.28	0 ± 0.00	10 ± 13.05
Staphylococcus aureus*	4 ± 0.54	34 ± 2.56	4 ± 1.00	14 ± 17.32	18 ± 1.50	2 ± 0.50	2 ± 0.50	7 ± 9.24
Sporolactobacillus sp.**	7 ± 1.50	0 ± 0.00	3 ± 1.00	3 ± 3.51	2 ± 0.50	10 ± 1.50	4 ± 0.50	5 ± 4.16
Klebsīlla pneumoniae*	2 ± 0.50	5 ± 0.52	0 ± 0.00	2 ± 2.52	1 ± 0.50	0 ± 0.00	0 ± 0.00	0 ± 0.00
Staphylococcus epidermidis*	4 ± 0.53	0 ± 0.00	0 ± 0.00	1 ± 2.31	8 ± 1.50	15 ± 1.26	0 ± 0.00	8 ± 7.51
Micrococcus luteus**	8 ± 2.30	0 ± 0.00	10 ± 2.24	6 ± 2.31	3 ± 0.58	11 ± 0.50	0 ± 0.00	5 ± 7.51
Total ml^{-1} ($\times 10^3$)	60 ± 4.39	174 ± 13.41	27 ± 2.99	87.00 ± 77.13	82 ± 7.74	206 ± 12.46	20 ± 2.57	102.67 ± 94.71

± = Standard Deviation
Unit for value is ml^{-1} ($\times 10^3$)

* = Pathogen
** = Non-Pathogen.

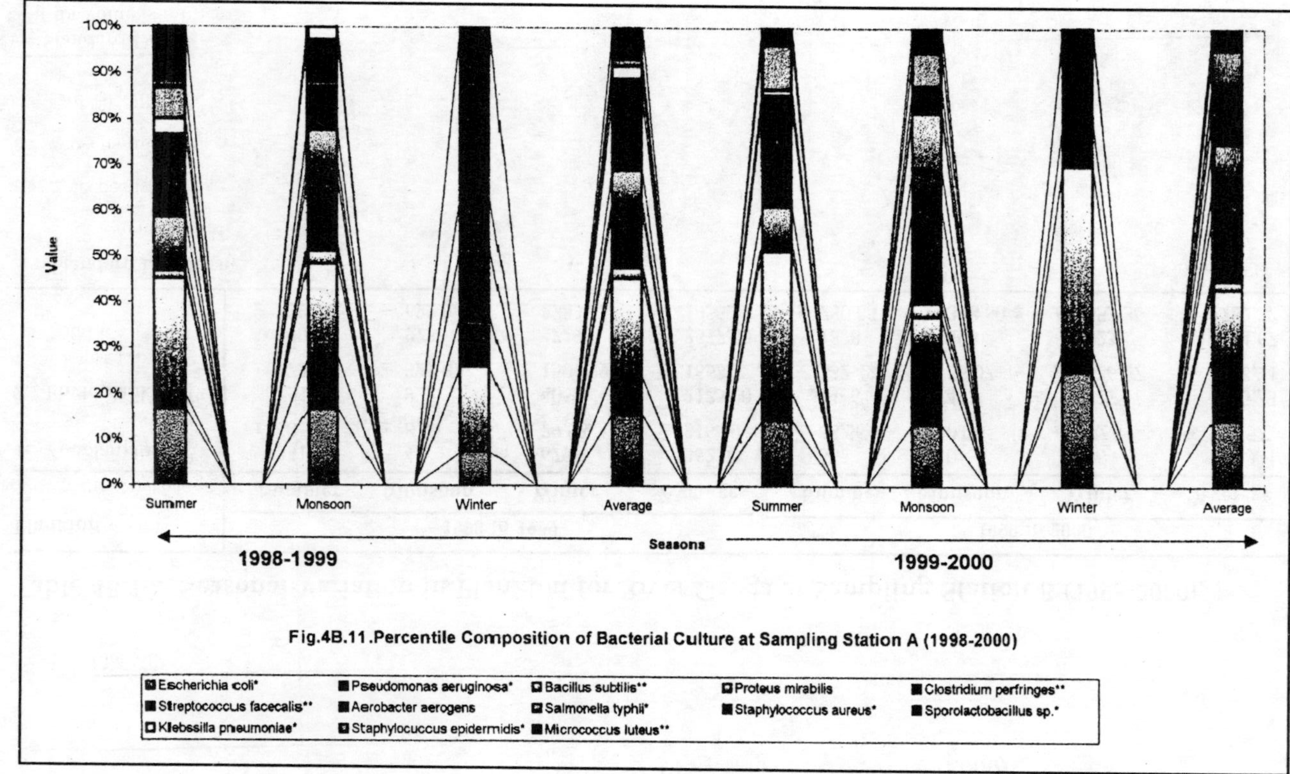

Fig.4B.11. Percentile Composition of Bacterial Culture at Sampling Station A (1998-2000)

Table 4B.10. Seasonal Variation in Plankton for River Ganga at Sampling Station B (1998-2000)

Plankton	1998 to 1999				1999 to 2000			
	Summer	Monsoon	Winter	Average	Summer	Monsoon	Winter	Average
1. Zooplankton l^{-1}	361 ± 60.38	46 ± 8.06	679 ± 79.58	362.00 ± 316.50	362 ± 68.56	0 ± 0.00	206 ± 13.63	189.33 ± 181.57
2. Phytoplankton l^{-1}	1548 ± 172.66	927 ± 321.67	4050 ± 1504.99	2175.00 ± 1653.22	1466 ± 332.42	493 ± 86.07	3817 ± 1536.07	1925.33 ± 1708.94
Total l^{-1}	1909 ± 839.34	973 ± 622.96	4729 ± 2383.66	2537.00 ± 1955.17	1828 ± 780.65	493 ± 348.60	4023 ± 2553.36	2114.67 ± 1782.37

± = Standard Deviation.

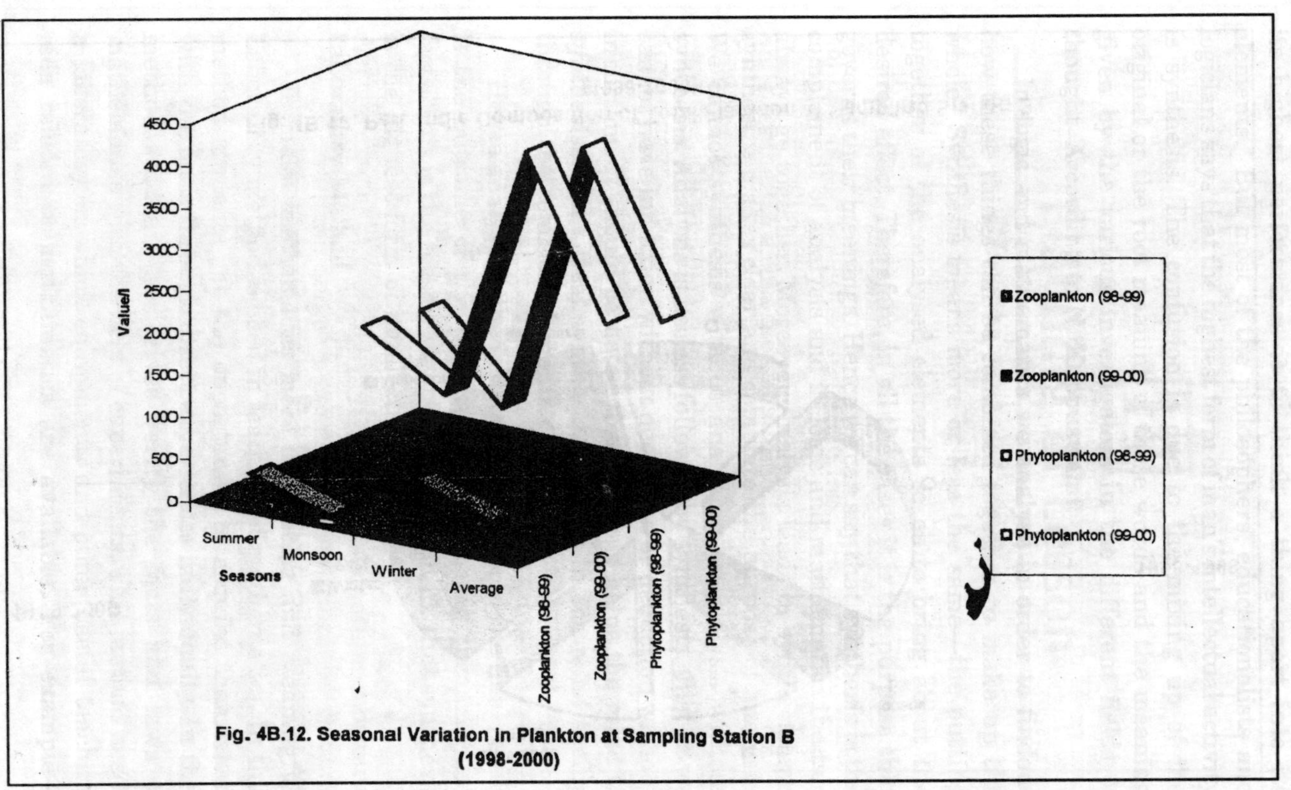

Fig. 4B.12. Seasonal Variation in Plankton at Sampling Station B (1998-2000)

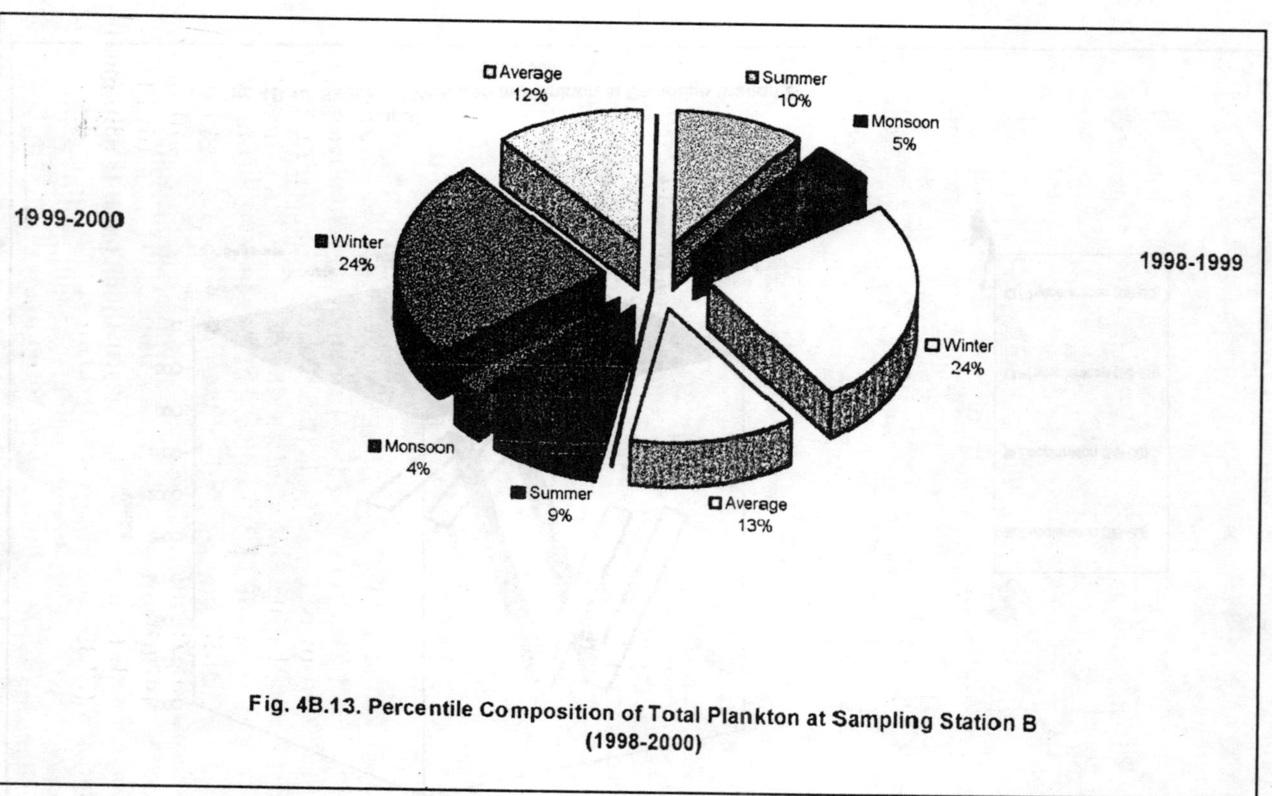

Fig. 4B.13. Percentile Composition of Total Plankton at Sampling Station B (1998-2000)

season of 1999-2000. In total at this sampling station winter season of 1998-1999 showed maximum (679 unit/l ±79.58) and minimum in monsoon season of 1999-2000 recorded as Nil. (Table 4B.11 and Fig. 4B.14).

The average value revealed that *Arcella* was dominating in the zooplankton 1998-1999 (56 unit/l ±52.85) followed by *Notholca* (54 unit/l ±48.69), *Philodima* (52 unit/l ±38.16), *Cyclops* (51 unit/l ±38.97), *Keratella* (47 unit/l ±46.01), *Amoeba* (39 unit/l ±41.20), *Daphnia* (32 unit/l ±28.01) and *Bosmina* (31 unit/l ±38.74). In second year 1999-2000 *Keratella* (37 unit/l ±44.11) found in strength followed by *Amoeba* (33 unit/l ±29.82), *Bosmina* (27 unit/l ±24.79), *Philodima* (25 unit/l ±23.35), *Daphnia* (22 unit/l ±22.50), *Arcella* (21 unit/l ±20.52), *Notholca* (13 unit/l ±20.23), *Cyclops* (11 unit/l ±18.19) (Table 4B.12 and Fig. 4B.15)

2. Phytoplankton

In phytoplankton **Bacillariophyceae** was dominating in both years and class **Myxophyceae** was found least in 1998-1999 and **Chlorophyceae** in 1999-2000. At this sampling station **Bacillariophyceae** was observed maximum in winter season (3016 unit/l ±80.32) and minimum in monsoon season (226 unit/l ±15.03) of 1999-2000. **Chlorophyceae** was found to be highest in winter season of 1998-1999 (1054 unit/l ±15.40) and lowest as Nil in monsoon season of 1998-1999. **Myxophyceae** showed optimum peak during winter season of 1999-2000 (682 unit/l ±109.09) and least peak during winter season of 1998-1999 (15 unit/l ±4.79) (Table 4B.13 and Figs. 4B.16, 4B.17).

After observing mean values it was noted that *Diatoms* was dominated (174.33 unit/l ±185.77) among **Bacillariophyceae** in 1998-1999 followed by *Cymbella* (150.67 unit/l ±142.47), *Tabellaria* (129.67 unit/l ±208.15), *Synedra* (128 unit/l ±197.54), *Surirella* (98.33 unit/l ±165.15), *Gomphonema* (63.33 unit/l ±74.27), *Cyclotella* (60.67 unit/l ±42.44), *Frustulia* (60 unit/l ±65.85), *Fragilaria* (55.33 unit/l ±77.78), *Phacus* (49.67 unit/l ±70.61), *Cocconeis* (47.67 unit/l ±79.12), *Amphora* (41 unit/l ±64.21), *Achnanthes* (39.67 unit/l ±59.28), *Navicula* (36.67 unit/l ±39.46), *Stauroneis* (27.33 unit/l ±38.80), *Pinnularia* (17.67 unit/l ±21.78) and *Nitzschia* (17 unit/l ±18.08). In second year 1999-2000, *Cymbella* (137.33 unit/l ±199.36) were recorded as a dominating

Table 4B.11. Quantitative Analysis for Zooplankton in Different Seasons of River Ganga at Sampling Station B (1998-2000)

	Seasons	Protozoa	Rotifera	Cladocera	Copepoda	Total l^{-1}
1998 to 1999	Summer	97 ± 20.51	173 ± 10.60	50 ± 32.53	41 ± 2.11	361 ± 60.38
	Monsoon	0 ± 0.00	12 ± 6.93	16 ± 11.31	18 ± 2.48	46 ± 8.06
	Winter	187 ± 16.26	274 ± 3.06	124 ± 18.38	94 ± 7.51	679 ± 79.58
	Average l^{-1}	94.67 ± 93.52	153.00 ± 132.14	63.33 ± 55.22	51.00 ± 38.97	362.00 ± 45.54
1999 to 2000	Summer	99 ± 12.02	168 ± 26.46	94 ± 2.83	1 ± 0.50	362 ± 68.56
	Monsoon	0 ± 0.00	0 ± 0.00	0 ± 0.00	0 ± 0.00	0 ± 0.00
	Winter	63 ± 13.44	58 ± 15.14	53 ± 6.36	32 ± 2.82	206 ± 13.63
	Average l^{-1}	54.00 ± 50.11	75.33 ± 85.33	49.00 ± 47.13	11.00 ± 18.19	189.33 ± 26.78

± = Standard Deviation.

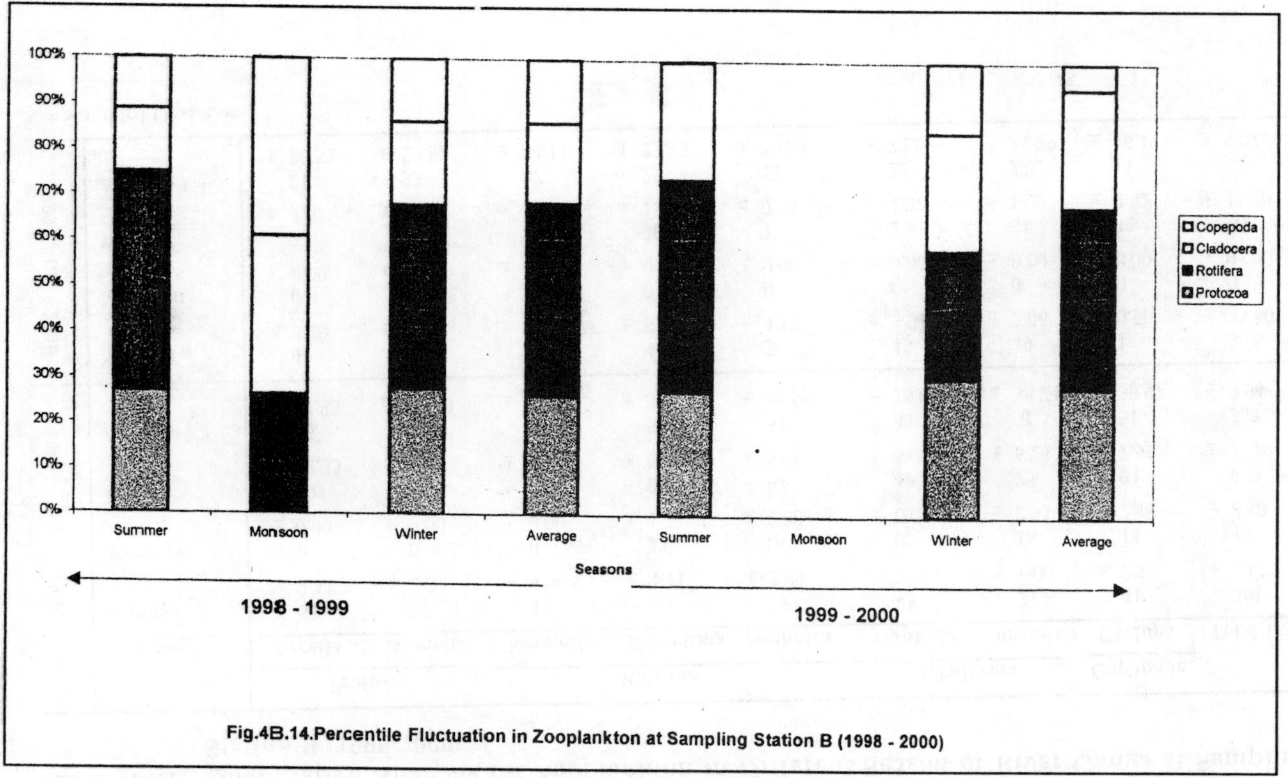

Fig.4B.14.Percentile Fluctuation in Zooplankton at Sampling Station B (1998 - 2000)

Table 4B.12. Quantitative Analysis for Zooplankton in Different Season of River Ganga at Sampling Station B (1998-2000)

		Protozoa		Rotifera			Cladocera		Copepoda	
	Seasons	Arcella	Amoeba	Keratella	Philodima	Notholca	Daphnia	Bosmina	Cyclops	Total l^{-1}
1998 to 1999	Summer	63 ± 9.33	34 ± 2.09	48 ± 4.06	56 ± 4.41	69 ± 3.89	48 ± 2.06	2 ± 1.00	41 ± 2.11	361 ± 20.79
	Monsoon	0 ± 0.00	0 ± 0.00	0 ± 0.00	12 ± 1.52	0 ± 0.00	0 ± 0.00	16 ± 2.51	18 ± 2.48	46 ± 8.10
	Winter	105 ± 12.53	82 ± 4.91	92 ± 4.44	88 ± 6.0[illegible]	94 ± 5.08	49 ± 6.06	75 ± 6.24	94 ± 7.51	679 ± 17.01
	Average l^{-1}	56 ± 52.85	39 ± 41.20	47 ± 46.01	52 ± 38.16	54 ± 48.69	32 ± 28.01	31 ± 38.74	51 ± 38.97	362 ± 9.94
1999 to 2000	Summer	41 ± 3.10	58 ± 5.02	86 ± 7.27	46 ± 3.82	36 ± 4.15	45 ± 3.51	49 ± 2.06	1 ± 0.50	362 ± 23.60
	Monsoon	0 ± 0.00	0 ± 0.00	0 ± 0.00	0 ± 0.00	0 ± 0.00	0 ± 0.00	0 ± 0.00	0 ± 0.00	0 ± 0.00
	Winter	22 ± 3.61	41 ± 2.31	26 ± 3.34	30 ± 1.06	2 ± 0.50	22 ± 4.22	31 ± 1.05	32 ± 2.82	206 ± 11.40
	Average l^{-1}	21 ± 20.52	33 ± 29.82	37 ± 44.11	25 ± 23.35	13 ± 20.23	22 ± 22.50	27 ± 24.79	11 ± 18.19	189 ± 9.07

± = Standard Deviation.

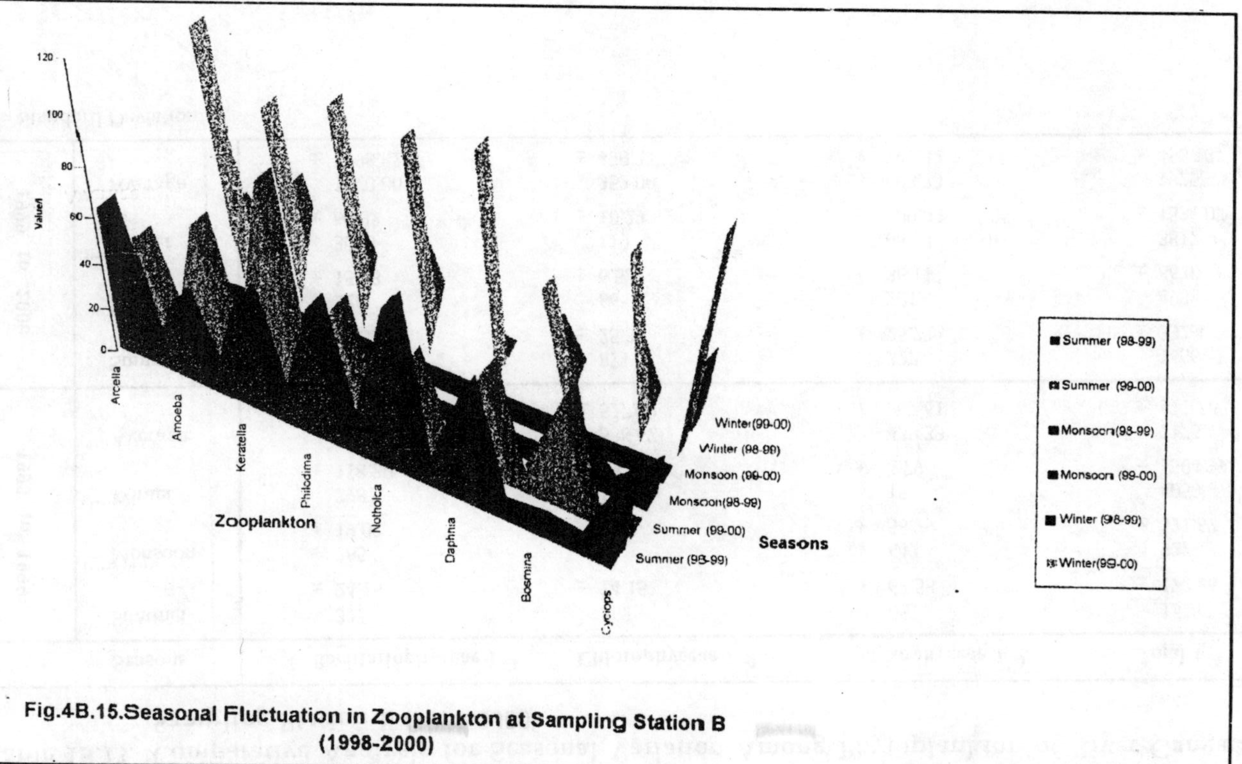

Fig.4B.15.Seasonal Fluctuation in Zooplankton at Sampling Station B (1998-2000)

Table 4B.13. Comparative Analysis for Seasonal Variation Among Phytoplankton of River Ganga at Sampling Station B (1998-2000)

	Seasons	Bacillariophyceae l^{-1}	Chlorophyceae l^{-1}	Myxophyceae l^{-1}	Total l^{-1}
1998 to 1999	Summer	325 ± 24.16	562 ± 14.15	661 ± 67.58	1548 ± 172.66
	Monsoon	285 ± 19.02	0 ± 0.00	642 ± 56.79	927 ± 321.67
	Winter	2981 ± 118.30	1054 ± 15.40	15 ± 4.79	4050 ± 1504.99
	Average	1197.00 ± 1545.12	538.67 ± 527.39	439.33 ± 367.61	2175 ± 411.77
1999 to 2000	Summer	268 ± 17.37	871 ± 25.34	327 ± 25.72	1466 ± 332.42
	Monsoon	226 ± 15.03	66 ± 6.53	201 ± 36.11	493 ± 86.07
	Winter	3016 ± 80.32	119 ± 10.29	682 ± 109.09	3817 ± 1536.07
	Average	1170.00 ± 1598.82	352.00 ± 450.25	403.33 ± 249.42	1925 ± 458.17

± = Standard Deviation.

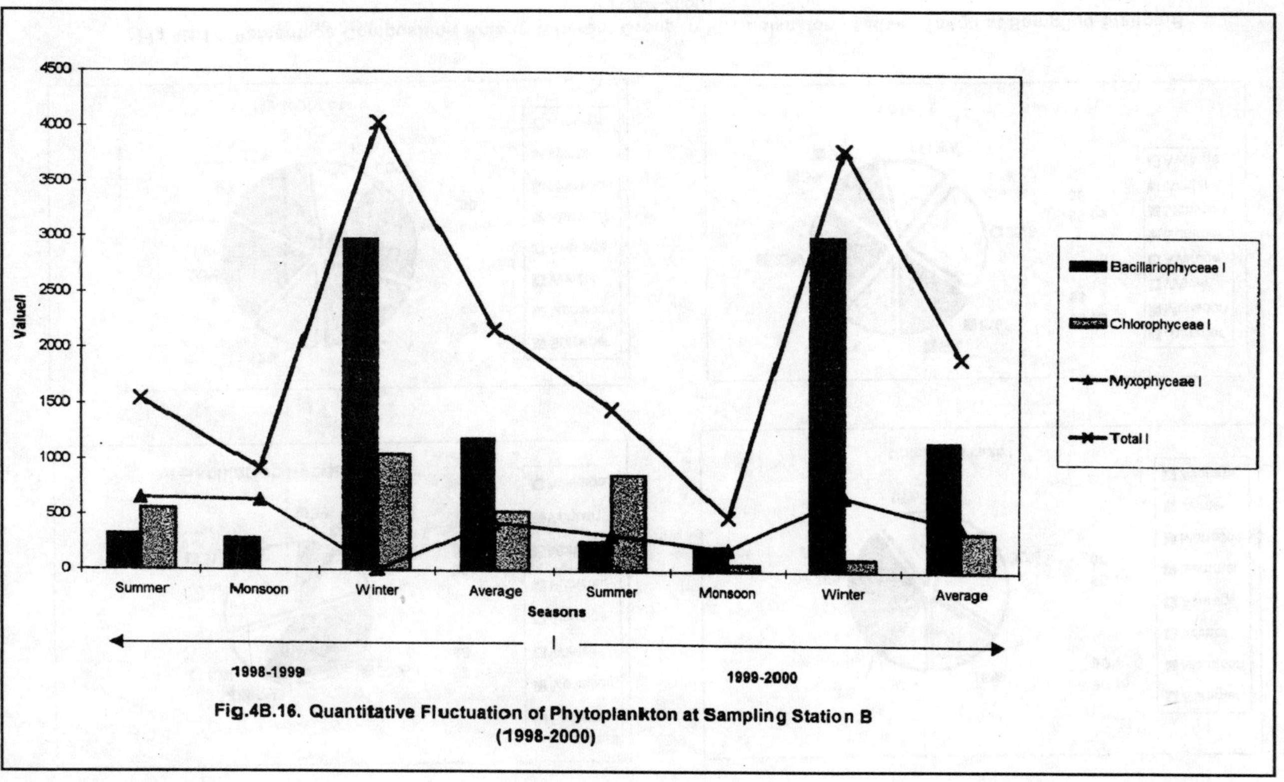

Fig.4B.16. Quantitative Fluctuation of Phytoplankton at Sampling Station B (1998-2000)

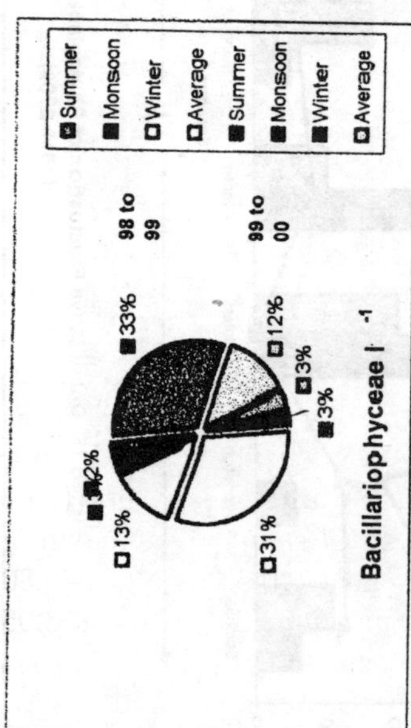

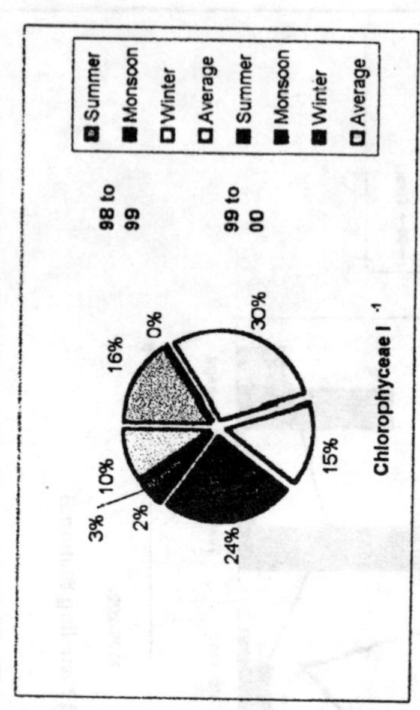

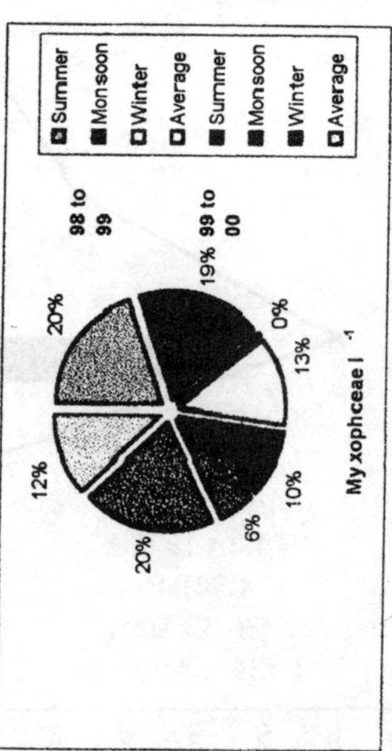

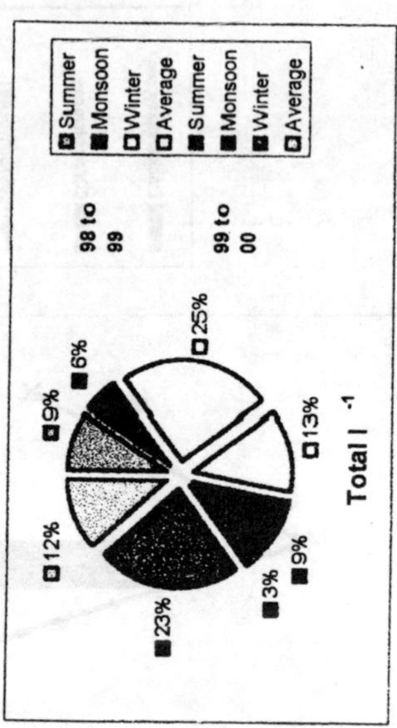

Fig 4B.17. Percentage Composition Among Different Group in Phytoplankton of River Ganga at Sampling Station B (1998-2000)

followed by *Diatoms* (121 unit/l ±149.94), *Surirella* (95.33 unit/l ±113.18), *Synedra* (85.33 unit/l ±108.74), *Frustulia* (82.33 unit/l ±91.00), *Tabellaria* (80 unit/l ±116.05), *Gomphonema* (76 unit/l ±104.85), *Achnanthes* (71.33 unit/l ±120.11), *Phacus* (69.67 unit/l ±118.94), *Amphora* (68 unit/l ±116.05), *Cyclops* (55 unit/l ±44.84), *Cocconeis* (55.33 unit/l ±90.65), *Stauroneis* (45.67 unit/l ±35.92), *Fragilaria* (40.33 unit/l ±53.14), *Navicula* (37 unit/l ±59.81), *Pinnularia* (29.67 unit/l ±51.38) and *Nitzschia* (22.67 unit/l ±31.66) (Table 4B.14 and Fig. 4B.18).

A great fluctuation among **Bacillariophyceae** was observed in seasonal study at this sampling point. In 1998-1999 summer, *Diatoms* was recorded highest (84 unit/l ±3.14) and *Nitzschia* and *Pinnularia* were found to be Nil; in monsoon season *Cymbella* was maximum (62 unit/l ±2.44) and *Amphora, Cocconeis* and *Surirella* were recorded to be Nil. In winter season *Diatoms* noted as optimum concentration (388 unit/l ±12.12) and least concentration was noted for *Nitzschia* (36 unit/l ±0.88). In 1999-2000 summer, *Diatoms* were recorded highest (59 unit/l ±5.26) and *Cocconeis, Navicula, Nitzschia, Pinnularia* and *Phacus* were recorded Nil; in monsoon season *Cyclotella* was noted maximum (45 unit/l ±0.86) while *Achnanthes, Amphora, Cymbella* and *Pinnularia* were recorded Nil. In winter season *Cymbella* was recorded in optimum concentration (266 unit/l ±14.19) and *Nitzschia* was observed least concentrated (59 unit/l ±6.06).

The *Cosmarium* observed to be highest average value (61.33 unit/l ±57.88) among **Chlorophyceae** in 1998-1999 followed by *Chaetophora* (54.67 unit/l ±54.01), *Schizogonium* (52.67 unit/l ±46.14), *Ulothrix* (52.67 unit/l ±51.08), *Oedogonium* (51.67 unit/l ±47.04), *Closterium* (51.33 unit/l ±51.00), *Spirogyra* (50 unit/l ±52.12), *Chlorella* (47.67 unit/l ±47.01), *Zygnema* (46.33 unit/l ±54.92), *Vaucheria* (41 unit/l ±42.58) and *Cladophora* (29.33 unit/l ±29.50). In 1999-2000, it revealed *Closterium* was maximum (46 unit/l ±74.54) followed by *Chlorella* (40.33 unit/l ±49.80), *Chaetophora* (36.33 unit/l ±56.04), *Spirogyra* (34 unit/l ±33.05), *Cosmarium* (34 unit/l ±58.89), *Vaucheria* (32 unit/l ±29.82), *Schizogonium* (28.67 unit/l ±23.67), *Cladophora* (27.33 unit/l ±37.43), *Zygnema* (26.33 unit/l ±38.89), *Oedogonium* (24 unit/l ±23.07) and *Ulothrix* (23 unit/l ±31.58).

Table. 4B.14. Quantitative Analysis for Seasonal Variation Among *Bacillariophyceae* of River Ganga at Sampling Station B (1998-2000)

	Seasons	Achnanthes	Amphora	Cocconeis	Cymbella	Cyclotella	Diatoms	Fragilaria	Frustulia	Gomphonena
1998 to 1999	Summer	9 ± 0.02	8 ± 0.04	4 ± 0.01	75 ± 2.44	26 ± 1.12	84 ± 3.14	6 ± 0.02	24 ± 0.09	24 ± 1.02
	Monsoon	2 ± 1.21	0 ± 0.00	0 ± 0.00	62 ± 2.44	48 ± 2.01	51 ± 1.04	15 ± 0.01	20 ± 0.04	17 ± 0.02
	Winter	108 ± 14.11	115 ± 12.10	139 ± 15.10	315 ± 15.11	108 ± 9.29	388 ± 12.12	145 ± 10.02	136 ± 9.16	149 ± 15.01
	Average l^{-1}	39.67 ± 59.28	41.00 ± 64.21	47.67 ± 79.12	150.67 ± 142.47	60.67 ± 42.44	174.33 ± 185.77	55.33 ± 77.78	60.00 ± 65.85	63.33 ± 74.27
1999 to 2000	Summer	4 ± 0.01	2 ± 0.01	0 ± 0.00	46 ± 10.15	16 ± 0.04	59 ± 5.26	18 ± 1.14	21 ± 1.28	19 ± 0.06
	Monsoon	0 ± 0.00	0 ± 0.00	2 ± 0.00	0 ± 0.00	45 ± 0.86	12 ± 0.07	2 ± 0.00	38 ± 3.06	12 ± 4.16
	Winter	210 ± 14.16	202 ± 13.11	158 ± 15.25	366 ± 14.19	104 ± 8.47	292 ± 12.11	101 ± 0.01	188 ± 2.12	197 ± 13.14
	Average l^{-1}	71.33 ± 120.11	68.00 ± 116.05	53.33 ± 90.65	137.33 ± 199.36	55.00 ± 44.84	121.00 ± 149.94	40.33 ± 53.14	82.33 ± 91.90	76.00 ± 104.85

Table 4B.14. (Contd.)

	Seasons	Navicula	Nitzschia	Pinnularia	Phacus	Stauroneis	Surirella	Synedra	Tabellaria	Total I $^{-1}$
1998 to 1999	Summer	10 ± 0.02	0 ± 0.00	0 ± 0.00	14 ± 0.01	8 ± 0.06	6 ± 0.01	20 ± 0.06	7 ± 0.01	325 ± 24.16
	Monsoon	18 ± 0.02	15 ± 0.03	11 ± 0.02	4 ± 0.03	2 ± 0.01	0 ± 0.00	8 ± 0.03	12 ± 0.02	285 ± 19.02
	Winter	82 ± 1.26	36 ± 0.88	42 ± 0.51	131 ± 5.42	72 ± 2.36	289 ± 11.41	356 ± 15.01	370 ± 14.07	2981 ± 118.30
	Average	36.67 ± 39.46	17.00 ± 18.08	17.67 ± 21.78	49.67 ± 70.61	27.33 ± 38.80	98.33 ± 165.15	128.00 ± 197.54	129.67 ± 208.15	1197.00 ± 47.90
1999 to 2000	Summer	0 ± 0.00	1 ± 0.00	0 ± 0.00	0 ± 0.00	28 ± 6.02	32 ± 0.05	10 ± 0.04	12 ± 0.53	268 ± 17.37
	Monsoon	5 ± 0.02	8 ± 0.01	0 ± 0.00	2 ± 0.00	22 ± 2.44	28 ± 5.16	36 ± 3.03	14 ± 0.06	226 ± 15.03
	Winter	106 ± 10.15	59 ± 6.06	89 ± 4.92	207 ± 13.15	87 ± 3.11	226 ± 14.56	210 ± 15.87	214 ± 14.62	3016 ± 80.32
	Average l $^{-1}$	37.00 ± 59.81	22.67 ± 31.66	29.67 ± 51.38	69.67 ± 118.94	45.67 ± 35.92	95.33 ± 113.18	85.33 ± 108.74	80.00 ± 116.05	1170.00 ± 30.84

± = Standard Deviation.

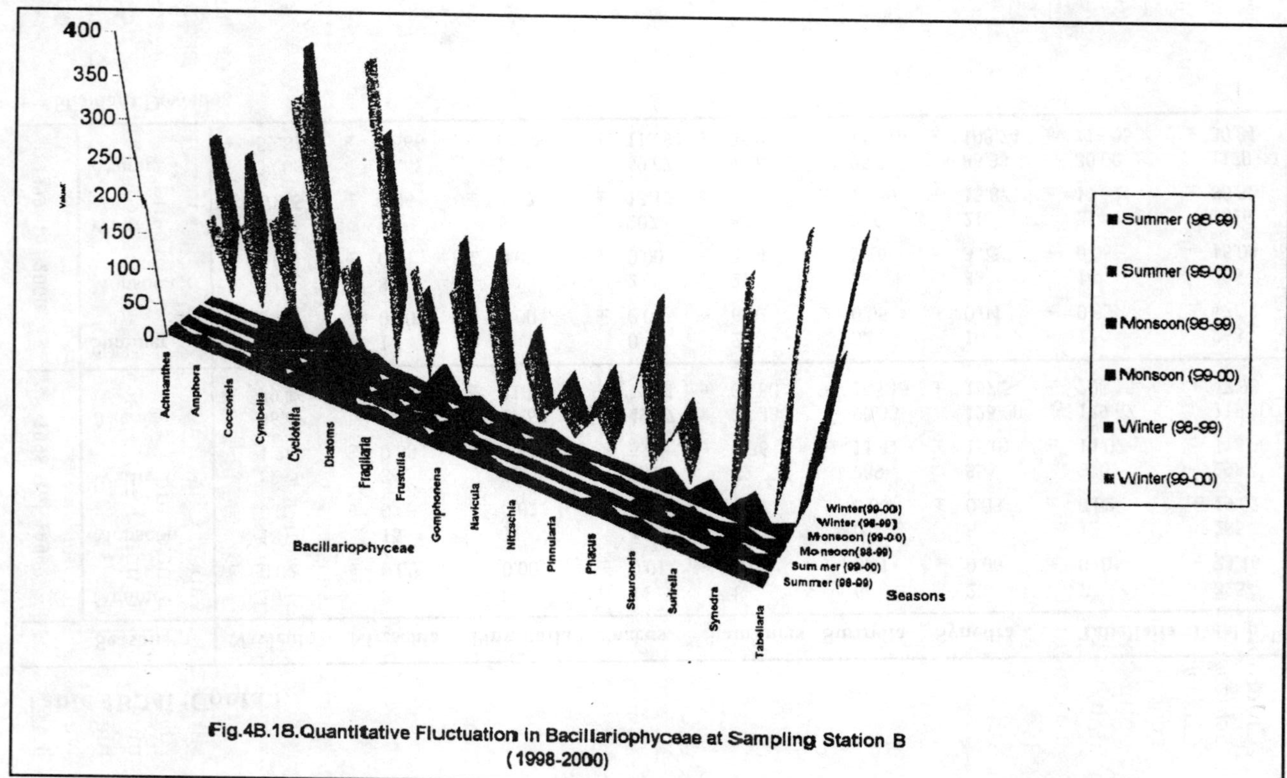

Fig.4B.18.Quantitative Fluctuation in Bacillariophyceae at Sampling Station B (1998-2000)

A great variation during different season was observed in **Chlorophyceae** at Har-ki-Pauri. In summer season 1998-1999, *Schizogonium* was recorded highest (72 unit/l ±5.46) and *Cladophora* was lowest (29 unit/l ±1.05). In monsoon season all genera was found Nil. In winter season *Cosmarium* was noted optimum (115 unit/l ±5.62) and Cladophora was least (59 unit/l ±4.12). In summer season 1999-2000, *Closterium* was noted maximum (132 unit/l ±3.26) and *Oedogonium* was minimum (48 unit/l ±5.24). In monsoon season *Spirogyra* was highest (18 unit/l ±2.61) and *Closterium, Cosmarium, Chlorella* and *Ulothrix* were found as Nil. In winter season Vaucheria was noted optimum (27 unit/l ±2.21) and *Cosmarium, Cladophora* and *Zygnema* was least as Nil. The total highest reading for **Chlorophyceae** was observed in winter (1054 unit/l ±15.40), and lowest in monsoon season as Nil for first year (1998-1999). In second year (1999-2000) maximum *Chlorophyceae* was detected during summer (871 unit/l ±25.34) and minimum in monsoon (66 unit/l ±6.53) (Table 4B.15, Fig. 4B.19).

In **Myxophyceae** revealing average value of *Anacystis* was noted optimum (148.67 unit/l ±131.15) in first year (1998-1999), followed by in descending order *Anabaena* (136.67 unit/l ±119.07), *Lyngbya* (85.67 unit/l ±65.68) and *Oscillatoria* (68.33 unit/l ±54.88). In second year (1999-2000) *Anacystis* was recorded highest (143.33 unit/l ±105.77), followed by *Oscillatoria* (107.33 unit/l ±140.23), *Anabaena* (93.33 unit/l ±6.43) and *Lyngbya* (59.33 unit/l ±4.16).

For seasonal fluctuation of **Myxophyceae** it is revealed that in summer of 1998-1999, *Anacystis* was existing in maximum (248 unit/l ±12.03) and *Oscillatoria* was minimum (102 unit/l ±12.44). In monsoon season *Anabaena* was observed to be highest (218 unit/l ±10.26), while *Oscillatoria* was lowest (98 unit/l ±5.49). In winter season *Lyngbya* was optimum (10 unit/l ±3.62) and *Anacystis* and *Anabaena* was found least as Nil. In second session of this study at Mayapuri Ghat it was recorded that in summer season *Anacystis* was maximum (109 unit/l ±10.01) and *Oscillatoria* was minimum (56 unit/l ±4.33), in monsoon season *Anabaena* was highest (86 unit/l ±6.22) and *Oscillatoria* was lowest as Nil, and in winter season *Oscillatoria* was optimum (266 unit/l ±14.19) and least was *Lyngbya* (58 unit/l ±15.25). In total value of 1998-2000

Table 4B.15. Quantitative Analysis for Seasonal Variation Among *chlorophyceae* of River Ganga at Sampling Station B (1998-2000)

	Seasons	Closterium	Cosmarium	Chaetophora	Chlorella	Cladophora	Oedogongonium
1998 to 1999	Summer	52 ± 2.31	69 ± 3.22	56 ± 1.06	49 ± 2.02	29 ± 1.05	63 ± 3.46
	Monsoon	0 ± 0.00	0 ± 0.00	0 ± 0.00	0 ± 0.00	0 ± 0.00	0 ± 0.00
	Winter	102 ± 10.33	115 ± 5.62	108 ± 10.41	94 ± 8.28	59 ± 4.12	92 ± 5.26
	Average I^{-1}	51.33 ± 51.00	61.33 ± 57.88	54.67 ± 54.01	47.67 ± 47.01	29.33 ± 29.50	51.67 ± 47.04
1999 to 2000	Summer	132 ± 3.26	102 ± 11.05	101 ± 10.25	96 ± 8.11	70 ± 6.23	48 ± 5.24
	Monsoon	0 ± 0.00	0 ± 0.00	6 ± 2.16	0 ± 0.00	12 ± 2.43	2 ± 0.57
	Winter	6 ± 0.54	0 ± 0.00	2 ± 1.15	25 ± 8.22	0 ± 0.00	22 ± 3.67
	Average I^{-1}	46.00 ± 74.54	34.00 ± 58.89	36.33 ± 56.04	40.33 ± 49.80	27.33 ± 37.43	24.00 ± 23.07

Table 4B.15. (Contd.)

	Seasons	Schizogonium	Spirogyra	Ulothrix	Vaucheria	Zygnema	Total I^{-1}
1998 to 1999	Summer	72 ± 5.46	46 ± 3.22	56 ± 1.98	38 ± 1.09	32 ± 3.15	562 ± 14.15
	Monsoon	0 ± 0.00	0 ± 0.00	0 ± 0.00	0 ± 0.00	0 ± 0.00	0 ± 0.00
	Winter	86 ± 6.05	104 ± 6.22	102 ± 11.37	85 ± 5.24	107 ± 10.24	1054 ± 15.40
	Average I^{-1}	52.67 ± 46.14	50.00 ± 52.12	52.67 ± 51.08	41.00 ± 42.58	46.33 ± 54.92	538.67 ± 8.27
1999 to 2000	Summer	56 ± 4.58	72 ± 3.88	59 ± 2.16	64 ± 3.24	71 ± 4.27	871 ± 25.34
	Monsoon	15 ± 3.24	18 ± 2.61	0 ± 0.00	5 ± 1.99	8 ± 1.01	66 ± 6.53
	Winter	15 ± 1.52	12 ± 5.16	10 ± 2.10	27 ± 2.21	0 ± 0.00	119 ± 10.29
	Average I^{-1}	28.67 ± 23.67	34.00 ± 33.05	23.00 ± 31.58	32.00 ± 29.82	26.33 ± 38.89	352.00 ± 7.09

± = Standard Deviation.

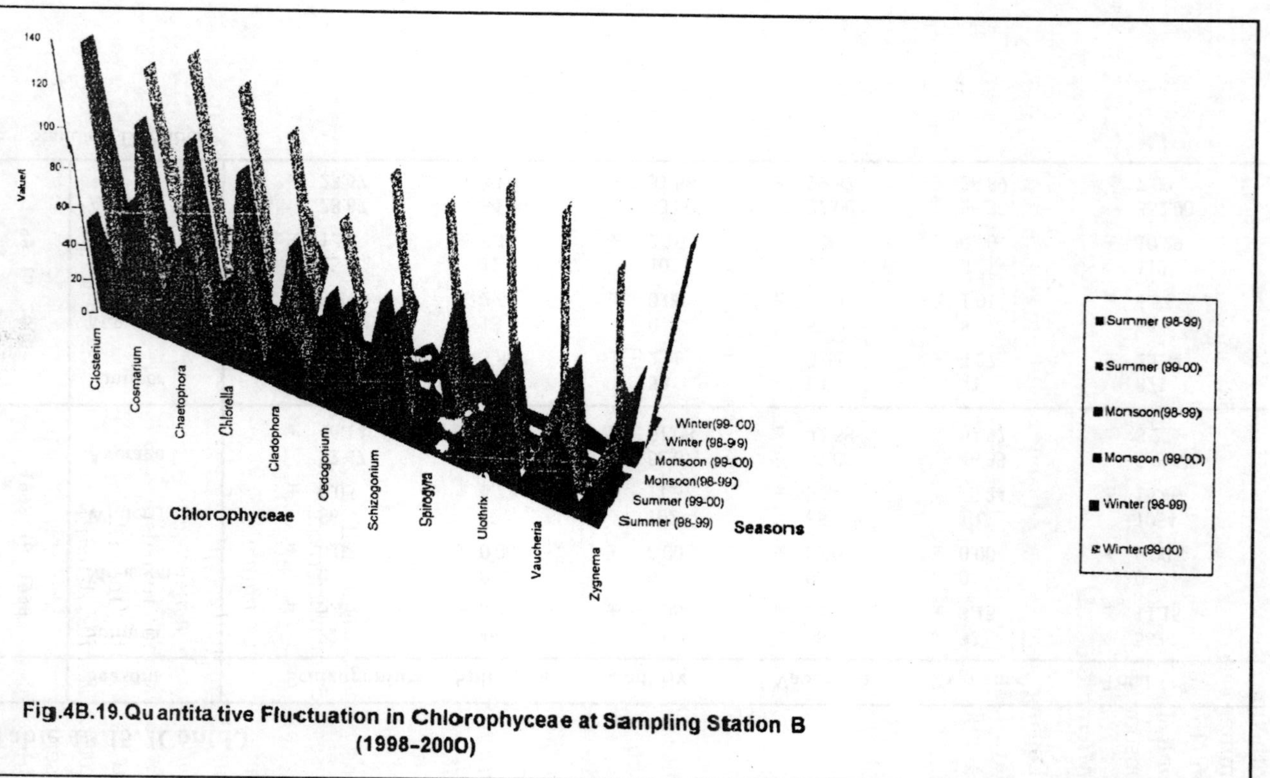

Fig.4B.19. Quantitative Fluctuation in Chlorophyceae at Sampling Station B (1998-2000)

study the maximum **Myxophyceae** for 1998-1999 were observed in summer season (661 unit/l ±67.58) and minimum in winter season (15 unit/l ±4.79), and in 1999-2000 highest recorded was (682 unit/l ±109.09) in winter and lowest was (201 unit/l ±36.11) in monsoon season (Table 4B.16, Fig. 4B.20).

(b) Microbiological

During analysis of Most Probable Number (MPN) and Standard Plate Count (SPC) at this sampling station observation revealed that mean MPN and SPC was higher (72.67 MPN/100ml ±58.48, 83.33x10^3 SPC/ml ±70.41) in 1999-200C and lowest in 1998-1999 (63.33 MPN/100 ml ±52.52, 55.67x10^3 SPC/ml ± 45.28). In the study of seasonal fluctuation, it is noted that in 1998-1999 monsoon was having MPN and SPC highest (115 MPN/100 ml ±11.31, 105.10^3 SPC/ml ±6.97) and in winter season lowest (10 MPN/100 ml ±6.05, 16x10^3 SPC/ml ±2.42). In 1999-2000 again monsoon season showed maximum MPN and SPC results (124 MPN/100 ml ±10.51, 159x10^3 SPC/ml ±9.34) and in winter season values were recorded as minimum (9 MPN/100 ml ±2.66, 20x10^3 SPC/ml ±2.03) (Table 4B.17 and Fig. 4B.21).

Bacterial cultures isolated from sampling station B shows that in 1998-1999 summer season, *Bacillus subtilis* was observed maximum (15x10^3 ml ±1.26) and minimum was *Pseudomonas aeruginosa, Proteus mirabilis, Clostridium perfringes, Streptococcus facecalis, Aerobacter aerogens* and *Salmonella typhii* as Nil. In monsoon season *Pseudomonas aeruginosa* was noted highest (22x10^3 ml ±2.56) and lowest was *Clostridium perfringes* and *Klebssilla pneumoniae* as Nil. In winter season optimum growth was observed for *Micrococcus luteus* (7x10^3/ml ±0.50) and least growth was. found to be Nil for *Escherichia coli, Pseudomonas aeruginosa, Proteus mirabilis, Clostridium perfringes, Aerobacter aerogens, Salmonella typhii, Sporolactobacillus sp., Klebssilla pneumoniae* and *Staphylococcus epidermidis*. In 1999-2000 summer season's observation revealed that growth of *Bacillus subtilis* was optimum (29x10^3/ml ±2.56) and *Clostridium perfringes, Streptococcus facecalis, Aerobacter aerogens* and *Klebssilla pneumoniae* was absent. In monsoon season growth of *Micrococcus luteus* was recorded as highest (29x10^3/ml ±1.54) and *Clostridium perfringes,*

Table 4B.16. Quantitative Analysis for Seasonal Variation Among *Myxophyceae* of River Ganga at Sampling Station B (1998-2000)

	Seasons	Anacystis	Anabaena	Lyngbya	Oscillatoria	Total l^{-1}
1998 to 1999	Summer	248 ± 12.03	192 ± 10.11	119 ± 13.51	102 ± 12.44	661 ± 67.58
	Monsoon	198 ± 15.61	218 ± 10.26	128 ± 9.15	98 ± 5.49	642 ± 56.79
	Winter	0 ± 0.00	0 ± 0.00	10 ± 3.62	5 ± 1.48	15 ± 4.79
	Average l^{-1}	148.67 ± 131.15	136.67 ± 119.07	85.67 ± 65.68	68.33 ± 54.88	439.33 ± 38.88
1999 to 2000	Summer	109 ± 10.01	98 ± 5.26	64 ± 4.21	56 ± 4.33	327 ± 25.72
	Monsoon	59 ± 5.31	86 ± 6.22	56 ± 2.49	0 ± 0.00	201 ± 36.11
	Winter	262 ± 14.16	96 ± 8.52	58 ± 15.25	266 ± 14.19	682 ± 109.09
	Average l^{-1}	143.33 ± 105.77	93.33 ± 6.43	59.33 ± 4.16	107.33 ± 140.23	403.33 ± 34.77

± = Standard Deviation.

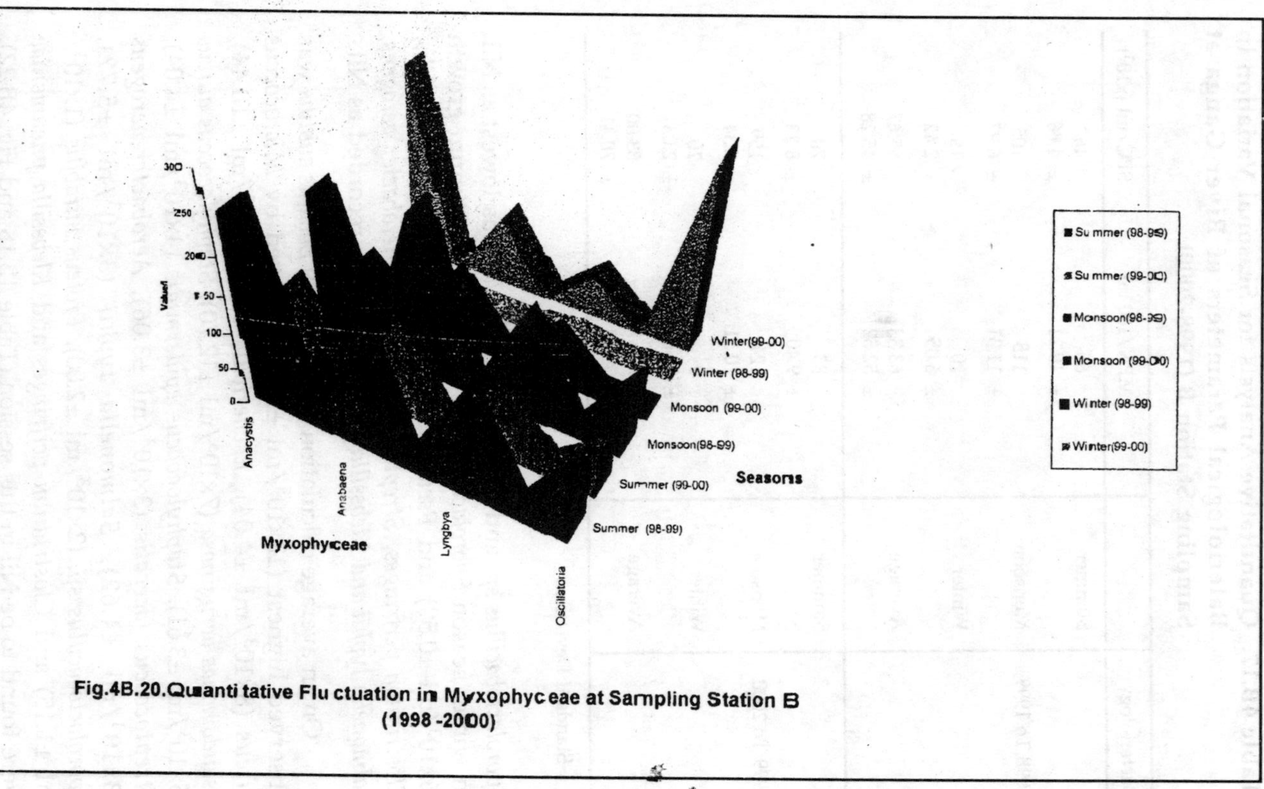

Fig.4B.20. Quantitative Fluctuation in Myxophyceae at Sampling Station B (1998-2000)

Table 4B.17. Quantitative Analysis for Seasonal Variation in Bateriological Parameters of River Ganga at Sampling Station B (1998-2000)

Bacteriology		MPN/100 ml	SPC/ml (x10³)
1998 To 1999	Summer	65 ± 10.11	46 ± 4.86
	Monsoon	115 ± 11.31	105 ± 6.97
	Winter	10 ± 6.05	16 ± 2.42
	Average	63.33 ± 52.52	55.67 ± 45.28
1999 To 2000	Summer	85 ± 9.40	70 ± 8.23
	Monsoon	124 ± 10.51	159 ± 9.34
	Winter	9 ± 2.66	20 ± 2.03
	Average	72.67 ± 58.48	83.00 ± 70.41

± = Standard Deviation.

Sporolactobacillus sp. and *Klebssilla pneumoniae* was lowest as Nil. In winter season *Sporolactobacillus sp.* showed optimum growth (6x10³/ml ±0.50) and *Pseudomonas aeruginosa, Proteus mirabilis, Clostridium perfringes, Streptococcus facecalis, Aerobacter aerogens, Salmonella typhii and Klebssilla pneumoniae* were recorded as Nil.

On an average calculation for 1998-1999 *Bacillus subtilis* was observed highest (12×10³/ml ±5.20) followed by *Micrococcus luteus* (8×10³/ml ±4.04), *Escherichia coli* (7×10³/ml ±10.44), *Pseudomonas aeruginosa* (7×10³/ml ±12.70), *Staphylococcus aureus* (5×10³/ml ±3.61), *Staphylococcus epidermidis* (4×10³/ml ±4.04), *Streptococcus facecalis* (3×10³/ml ±3.06), *Aerobacter aerogens* (3x10³/ml ±4.62), *Salmonella typhii* (3×10³/ml ±5.77), *Sporolactobacillus sp.* (2×10³/ml ±2.00), *Proteus mirabilis* (1×10³/ml ±1.15) and *Clostridium perfringes* and *Klebssilla pneumoniae* were found to be Nil in this session (Table 4B.18 and Fig. 4B.22).

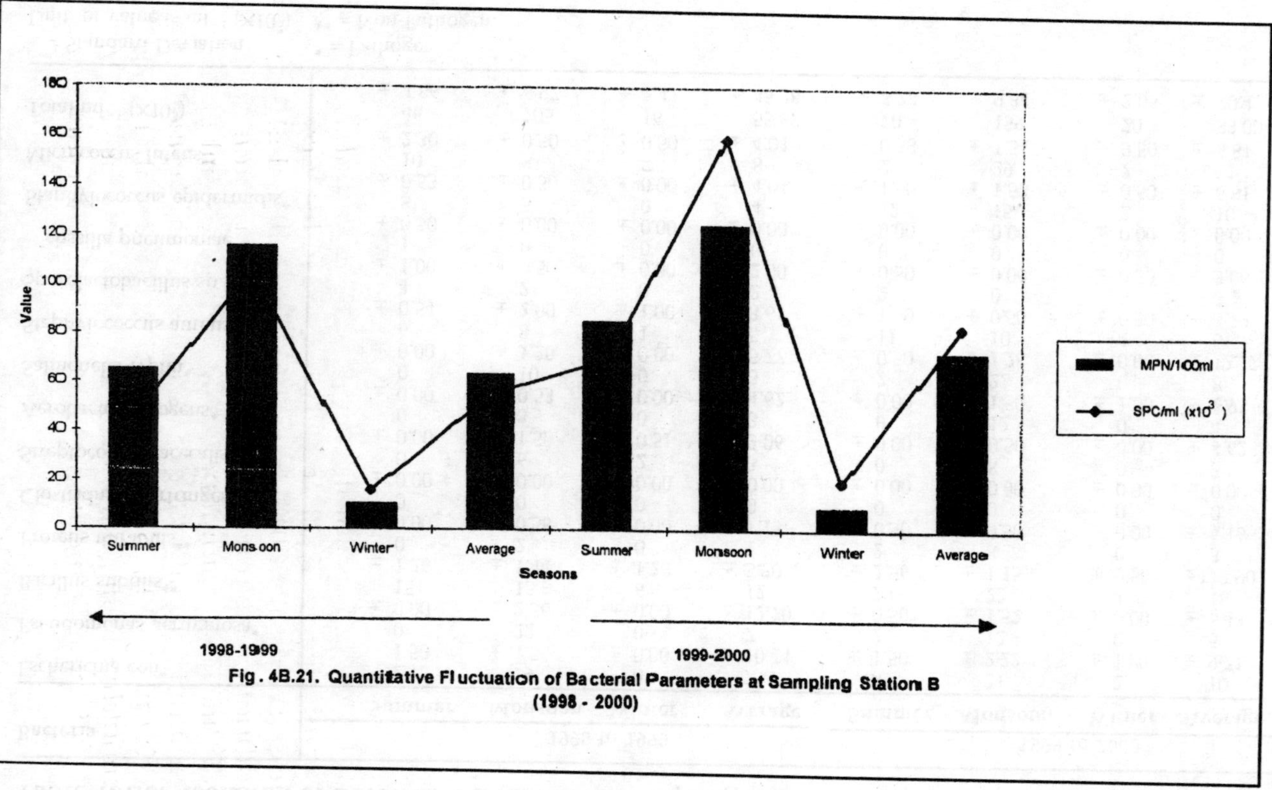

Fig. 4B.21. Quantitative Fluctuation of Bacterial Parameters at Sampling Station B (1998 - 2000)

Table 4B.18. Isolation of Bacterial Culture from Sampling Station B (1998-2000)

Bacteria	1998 to 1999				1999 to 2000			
	Summer	Monsoon	Winter	Average	Summer	Monsoon	Winter	Average
Escherichia coli*	2 ± 1.50	19 ± 1.56	0 ± 0.00	7 ± 10.44	8 ± 1.50	21 ± 2.22	2 ± 1.00	10 ± 9.71
Pseudomonas aeruginosa*	0 ± 0.00	22 ± 2.56	0 ± 0.00	7 ± 12.70	2 ± 0.50	12 ± 1.52	0 ± 0.00	5 ± 6.43
Bacillus subtilis**	15 ± 1.26	15 ± 1.44	6 ± 1.20	12 ± 5.20	29 ± 2.56	22 ± 1.15	4 ± 0.50	18 ± 12.90
Proteus mirabilis**	0 ± 0.00	2 ± 0.56	0 ± 0.00	1 ± 1.15	2 ± 0.50	8 ± 0.50	0 ± 0.00	3 ± 4.16
Clostridium perfringes*	0 ± 0.00	0 ± 0.00	0 ± 0.00	0 ± 0.00	0 ± 0.00	0 ± 0.00	0 ± 0.00	0 ± 0.00
Streptococcus facecalis*	0 ± 0.00	6 ± 1.50	2 ± 0.51	3 ± 3.06	0 ± 0.00	8 ± 0.50	0 ± 0.00	3 ± 4.62
Aerobacter aerogens*	0 ± 0.00	8 ± 0.53	0 ± 0.00	3 ± 4.62	0 ± 0.00	12 ± 1.50	0 ± 1.00	4 ± 6.93
Salmonella typhii*	0 ± 0.00	10 ± 1.20	0 ± 0.00	3 ± 5.77	2 ± 0.50	22 ± 1.28	0 ± 0.00	8 ± 12.17
Staphylococcus aureus*	6 ± 0.54	8 ± 2.00	1 ± 1.00	5 ± 3.61	11 ± 1.50	10 ± 0.50	4 ± 0.50	8 ± 3.79
Sporolactobacillus sp.**	4 ± 1.00	2 ± 0.50	0 ± 0.00	2 ± 2.00	2 ± 0.50	0 ± 0.00	6 ± 0.50	3 ± 3.06
[illegible]ebssilla pneumoniae*	1 ± 0.50	0 ± 0.00	0 ± 0.00	0 ± 0.00	0 ± 0.00	0 ± 0.00	0 ± 0.00	0 ± 0.00
Staphylococcus epidermidis*	8 ± 0.53	5 ± 0.50	0 ± 0.00	4 ± 4.04	12 ± 1.50	15 ± 1.58	2 ± 0.50	10 ± 6.81
Micrococcus luteus**	10 ± 2.30	8 ± 0.50	7 ± 0.50	8 ± 4.04	2 ± 0.58	29 ± 1.54	2 ± 0.50	11 ± 6.81
Total ml $^{-1}$ ($\times 10^3$)	46 ± 4.86	105 ± 6.97	16 ± 2.42	55.67 ± 45.28	70 ± 8.23	159 ± 9.34	20 ± 2.03	83.00 ± 70.41

± = Standard Deviation   * = Pathogen

Unit for value is ml $^{-1}$ ($\times 10^3$)   ** = Non-Pathogen.

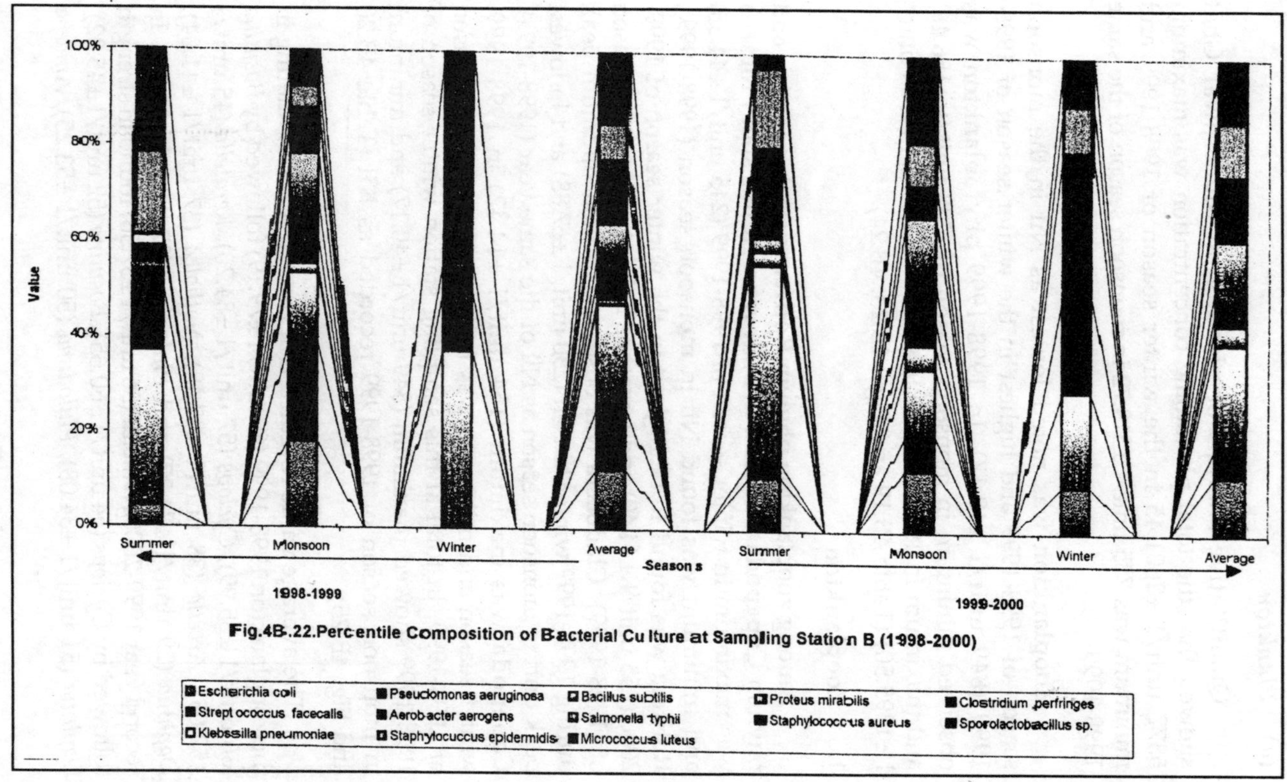

Fig.4B.22.Percentile Composition of Bacterial Culture at Sampling Station B (1998-2000)

SAMPLING STATION C (Muneshwar Ghat)

(a) Plankton

Quantitative analysis of plankton at Muneshwar Ghat showed that the total planktonic concentration was maximum 5377 unit/l ±2601.45 in the winter season of 1998-1999 and minimum was 735 unit/l ±519.72 in monsoon season of the same 1998-1999.

Zooplankton was noted lowest as Nil in the monsoon season of 1998-1999 and highest in the winter season of 1998-1999 (849 unit/l ±40.17). In 1998-1999 phytoplankton was observed minimum in monsoon season of (735 unit/l ±156.46) and maximum in 1998-1999 for the season of winter (4528 unit/l ±1395.50) (Table 4B.19 and Figs. 4B.23, 4B.24).

1. Zooplankton

Among zooplankton phylum **Rotifera** was dominating and phylum **Copepoda** was least in both year. At this spot **Protozoa** was maximum in winter season of 1998-1999 (245 unit/l ±44.55) and minimum was found Nil in monsoon season (1998-1999). **Rotifera** was found to be highest in the winter season of 1999-2000 (255 unit/l ±44.40) and lowest as Nil in the monsoon season of (1998-1999). **Cladocera** was observed in their optimum peak during 1998-1999 winter season (230 unit/l ±67.88) and in lowest peak during monsoon season as Nil of the same year (1998-1999). **Copepoda** was maximum (154 unit/l ±14.42) in 1998-1999, winter season and minimum was found Nil in monsoon season of 1998-1999. In total at this sampling station winter season of 1998-1999 showed maximum (849 unit/l ±40.17) and minimum in monsoon season of 1998-1999 recorded as Nil (Table 4B.20 and Fig. 4B.25).

The average value revealed that *Amoeba* was dominating in the zooplankton 1998-1999 (66 unit/l ±79.19) followed by *Bosmina* (66 unit/l ±85.99), *Cyclops* (57 unit/l ±84.20), *Keratella* (45 unit/l ±52.05), *Arcella* (38 unit/l ±47.48), *Notholca* (37 unit/l ±41.04), *Daphnia* (29 unit/l ±34.27) and *Philodima* (27 unit/l ±23.64). In second year 1999-2000 *Notholca* (56 unit/l ±57.56) found in strength followed by *Cyclops* (54 unit/l ±60.65), *Bosmina* (52 unit/l ±45.32), *Daphnia* (51 unit/l ±34.08), *Philodima* (50 unit/l ±53.25), *Amoeba*

Table 4B. 19. Seasonal Variation in Plankton for River Ganga at Sampling Station C (1998-2000)

Plankton	1998 To 1999				1999 to 2000			
	Summer	Monsoon	Winter	Average	Summer	Monsoon	Winter	Average
1. Zooplankton l^{-1}	246	0	849	365.00	367	94	657	372.67
	± 36.26	± 0.00	± 40.17	± 436.83	± 66.79	± 13.38	± 67.28	± 281.54
2. Phytoplankton l^{-1}	1508	735	4528	2257.00	2592	1034	3968	2531.33
	± 170.89	± 156.46	± 1395.50	± 2004.36	± 639.62	± 529.59	± 1584.23	± 1467.94
Total l^{-1}	1754	735	5377	2622.00	2959	1128	4625	2904.00
	± 892.37	± 519.72	± 2601.45	± 2439.69	± 1573.31	± 664.68	± 2341.23	± 1749.15

± = Standard Deviation.

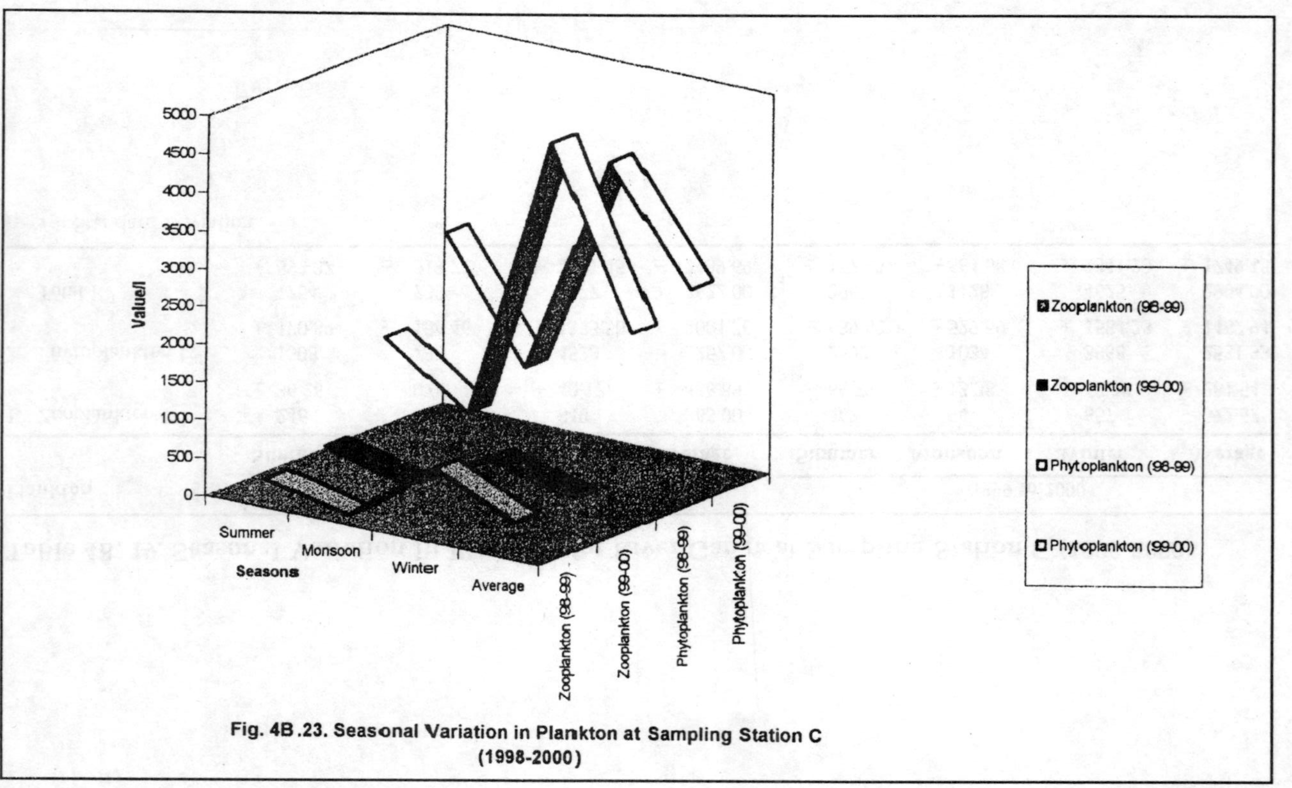

Fig. 4B.23. Seasonal Variation in Plankton at Sampling Station C (1998-2000)

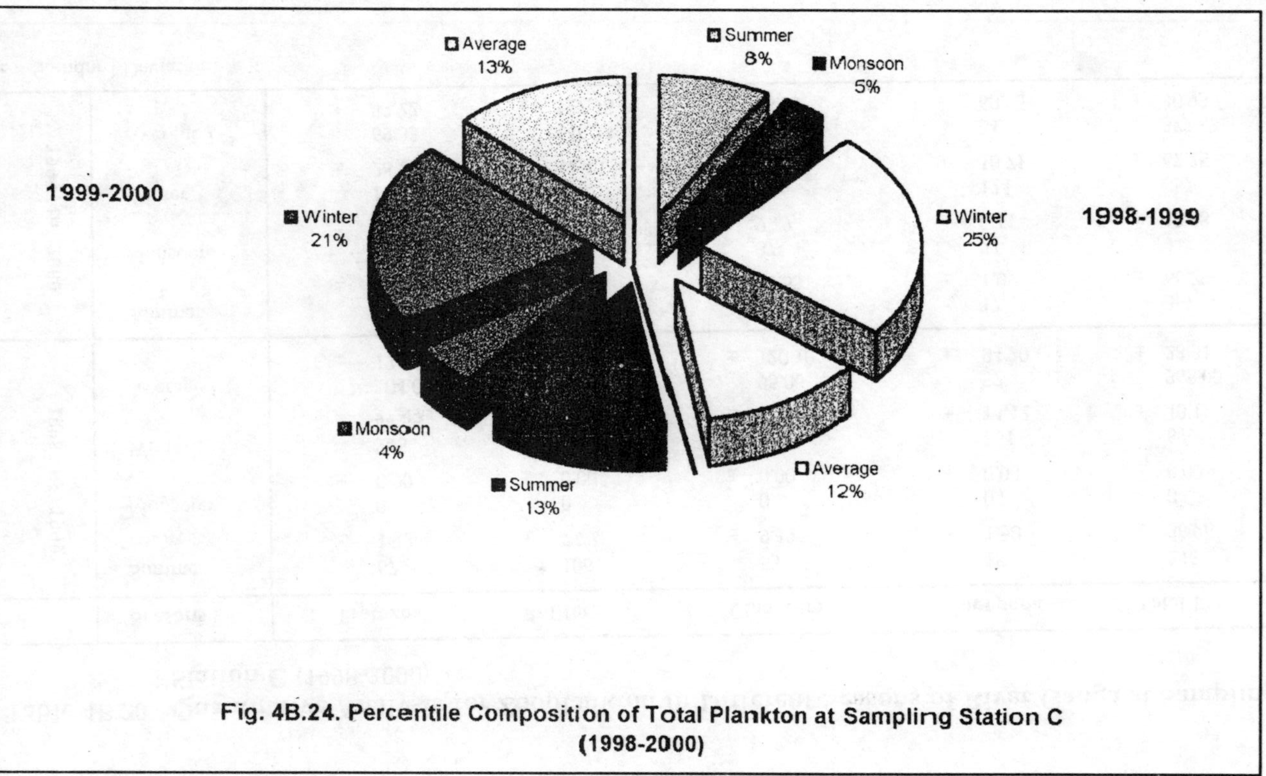

Fig. 4B.24. Percentile Composition of Total Plankton at Sampling Station C (1998-2000)

Table 4B.20. Quantitative Analysis for Zooplankton in Different Seasons of River Ganga at Sampling Station C (1998-2000)

	Seasons	Protozoa	Rotifera	Cladocera	Copepoda	Total I $^{-1}$
1998 to 1999	Summer	67 ± 16.26	106 ± 7.77	55 ± 9.19	18 ± 1.50	246 ± 36.26
	Monsoon	0 ± 0.00	0 ± 0.00	0 ± 0.00	0 ± 0.00	0 ± 0.00
	Winter	245 ± 44.55	220 ± 33.17	230 ± 67.88	154 ± 14.42	849 ± 40.17
	Average I $^{-1}$	104.00 ± 126.62	108.67 ± 110.02	95.00 ± 120.10	57 ± 84.20	365.00 ± 23.31
1999 to 2000	Summer	74 ± 7.07	178 ± 19.30	98 ± 32.53	17 ± 1.09	367 ± 66.79
	Monsoon	30 ± 5.66	6 ± 3.46	37 ± 9.19	21 ± 2.11	94 ± 13.38
	Winter	104 ± 28.28	255 ± 44.40	174 ± 24.04	124 ± 10.71	657 ± 67.28
	Average I $^{-1}$	69.33 ± 37.22	146.33 ± 127.48	103.00 ± 68.64	54 ± 60.65	372.67 ± 40.93

± = Standard Deviation.

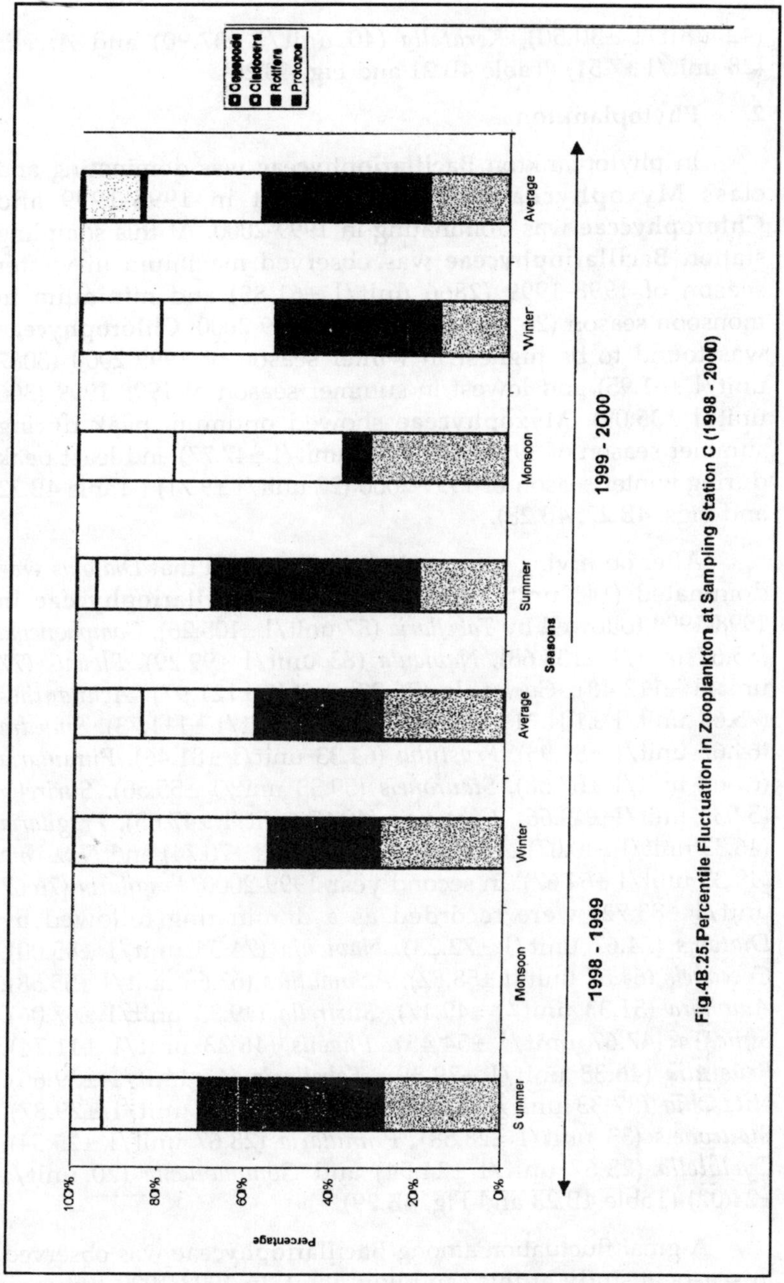

Fig.4B.25.Percentile Fluctuation in Zooplankton at Sampling Station C (1998 - 2000)

(42 unit/l ±30.50), *Keratella* (40 unit/l ±37.90) and *Arcella* (28 unit/l ±7.51) (Table 4B.21 and Fig. 4B.26).

2. Phytoplankton

In phytoplankton **Bacillariophyceae** was dominating and class **Myxophyceae** was found least in 1998-1999 and **Chlorophyceae** was dominating in 1999-2000. At this sampling station **Bacillariophyceae** was observed maximum in winter season of 1998-1999 (2866 unit/l ±61.89) and minimum in monsoon season (26 unit/l ±4.29) of 1999-2000. **Chlorophyceae** was found to be highest in winter season of 1999-2000 (3087 unit/l ±61.95) and lowest in summer season of 1998-1999 (306 unit/l ±35.09). **Myxophyceae** showed optimum peak during summer season of 1998-1999 (615 unit/l ±47.77) and least peak during winter season of 1999-2000 (22 unit/l ±9.71) (Table 4B.22 and Figs. 4B.27, 4B.28).

After observing mean values it was noted that *Diatoms* was dominated (141 unit/l ±153.84) among **Bacillariophyceae** in 1998-1999 followed by *Tabellaria* (87 unit/l ±105.26), *Gomphonema* (85.67 unit/l ±133.68), *Navicula* (83 unit/l ±99.29), *Phacus* (77 unit/l ±113.48), *Cyclotella* (76.33 unit/l ±121.97), *Achnanthes* (75.67 unit/l ±104.21), *Cymbella* (75.67 unit/l ±111.73), *Synedra* (64.67 unit/l ±87.96), *Frustulia* (64.33 unit/l ±81.46), *Pinnularia* (61.67 unit/l ±67.68), *Stauroneis* (59.33 unit/l ±55.50), *Surirella* (57.33 unit/l ±49.66), *Cocconeis* (49.67 unit/l ±47.65), *Fragilaria* (46.33 unit/l ±40.07), *Amphora* (42.33 unit/l ±70.74) and *Nitzschia* (40.33 unit/l ±64.67). In second year 1999-2000, *Fragilaria* (76.67 unit/l ±83.72) were recorded as a dominating followed by *Diatoms* (74.67 unit/l ±72.23), *Navicula* (74.33 unit/l ±85.00), *Cocconeis* (64.67 unit/l ±58.82), *Achnanthes* (63.66 unit/l ±55.58), *Amphora* (51.33 unit/l ±49.17), *Surirella* (49.33 unit/l ±47.06), *Synedra* (47.67 unit/l ±54.45), *Phacus* (46.33 unit/l ±41.74), *Frustulia* (46.33 unit/l ±79.39), *Tabellaria* (45 unit/l ±39.66), *Nitzschia* (37.33 unit/l ±43.66), *Cymbella* (34 unit/l ±29.87), *Stauroneis* (33 unit/l ±28.58), *Pinnularia* (28.67 unit/l ±25.54), *Cyclotella* (23.67 unit/l ±24.54) and *Gomphonema* (20 unit/l ±24.02) (Table 4B.23 and Fig. 4B.29).

A great fluctuation among **Bacillariophyceae** was observed in seasonal study at this sampling point. In 1998-1999 summer,

Table. 4B.21. Quantitative Analysis for Zooplankton in Different Season of River Ganga at Sampling Station C (1998-2000)

		Protozoa		Rotifera			Cladocera		Copepoda	
	Seasons	Arcella	Amoeba	Keratella	Philodima	Notholca	Daphnia	Bosmina	Cyclops	Total l^{-1}
1998 to 1999	Summer	22 ± 3.24	45 ± 4.19	33 ± 2.09	44 ± 3.16	29 ± 1.08	21 ± 2.06	34 ± 1.06	18 ± 1.50	246 ± 10.22
	Monsoon	0 ± 0.00	0 ± 0.00	0 ± 0.00	0 ± 0.00	0 ± 0.00	0 ± 0.00	0 ± 0.00	0 ± 0.00	0 ± 0.00
	Winter	91 ± 6.81	154 ± 12.48	102 ± 12.51	37 ± 4.28	81 ± 3.18	67 ± 2.41	163 ± 10.63	154 ± 14.42	849 ± 46.30
	Average l^{-1}	38 ± 47.48	66 ± 79.19	45 ± 52.05	27 ± 23.64	37 ± 41.04	29 ± 34.27	66 ± 85.99	57 ± 84.20	365 ± 15.69
1999 to 2000	Summer	32 ± 3.22	42 ± 5.16	81 ± 4.62	44 ± 3.27	53 ± 6.27	72 ± 3.34	26 ± 2.50	17 ± 1.09	367 ± 22.06
	Monsoon	19 ± 1.50	11 ± 0.50	6 ± 1.05	0 ± 0.00	0 ± 0.00	12 ± 2.43	25 ± 2.04	21 ± 2.11	94 ± 9.44
	Winter	32 ± 2.55	72 ± 5.12	34 ± 3.16	106 ± 11.42	115 ± 10.58	70 ± 5.02	104 ± 9.27	124 ± 10.71	657 ± 35.78
	Average l^{-1}	28 ± 7.51	42 ± 30.50	40 ± 37.90	50 ± 53.25	56 ± 57.56	51 ± 34.08	52 ± 45.32	54 ± 60.65	373 ± 9.45

± = Standard Deviation.

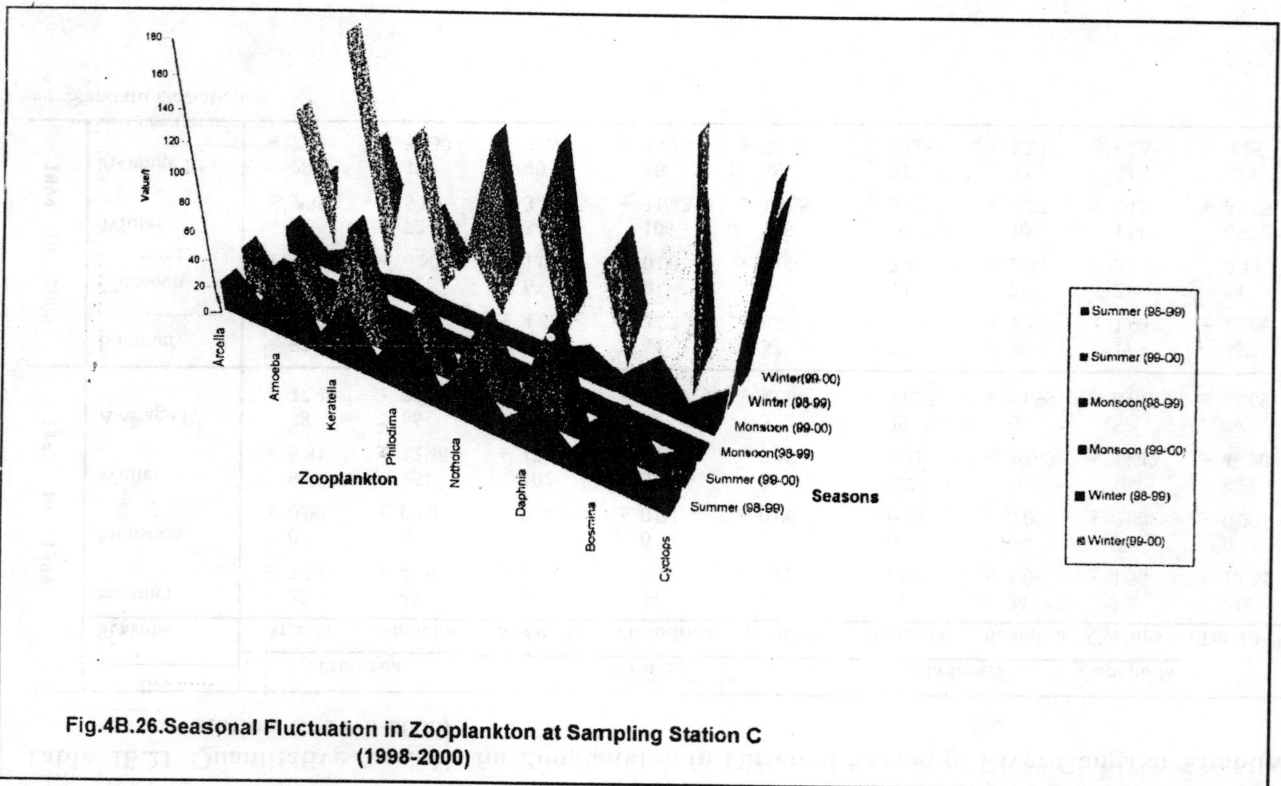

Fig.4B.26.Seasonal Fluctuation in Zooplankton at Sampling Station C (1998-2000)

Table 4B.22. Comparative Analysis for Seasonal Variation Among Phytoplankton of River Ganga at Sampling Station C (1998-2000)

	Seasons	Bacillariophyceae l^{-1}	Chlorophyceae l^{-1}	Myxophyceae l^{-1}	Total l^{-1}
1998 to 1999	Summer	587 ± 27.04	306 ± 35.09	615 ± 47.77	1508 ± 170.89
	Monsoon	109 ± 6.59	416 ± 29.11	210 ± 18.81	735 ± 156.46
	Winter	2866 ± 61.89	1584 ± 57.45	78 ± 5.07	4528 ± 1395.50
	Average	1187.33 ± 1473.28	768.67 ± 708.24	301.00 ± 279.83	2257 ± 443.39
1999 to 2000	Summer	1564 ± 43.07	718 ± 45.22	310 ± 11.47	2592 ± 639.62
	Monsoon	26 ± 4.29	956 ± 48.39	52 ± 8.91	1034 ± 529.59
	Winter	859 ± 28.23	3087 ± 61.95	22 ± 9.71	3968 ± 1584.23
	Average	816.33 ± 769.89	1587.00 ± 1304.48	128.00 ± 158.33	2531 ± 729.89

± = Standard Deviation.

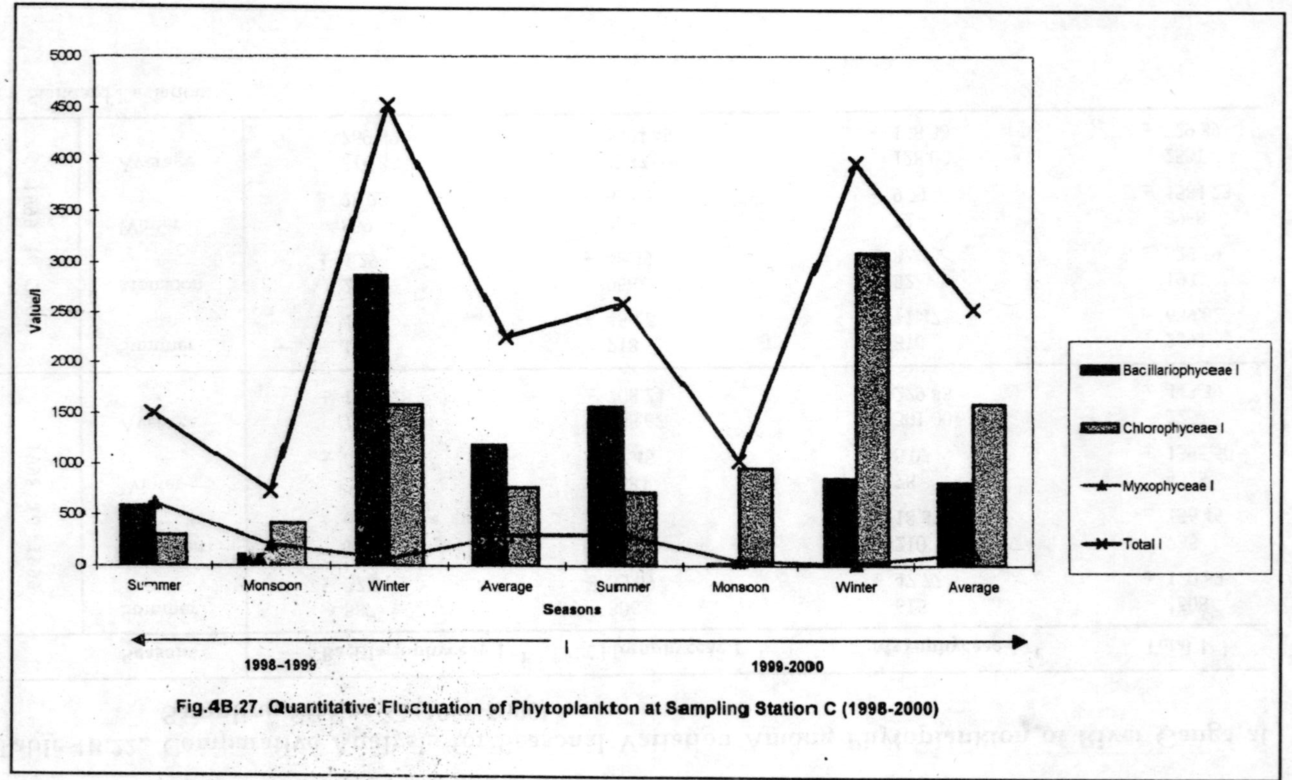

Fig.4B.27. Quantitative Fluctuation of Phytoplankton at Sampling Station C (1998-2000)

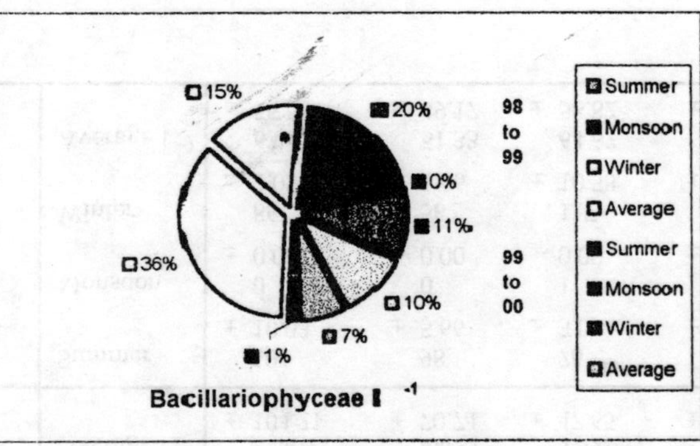

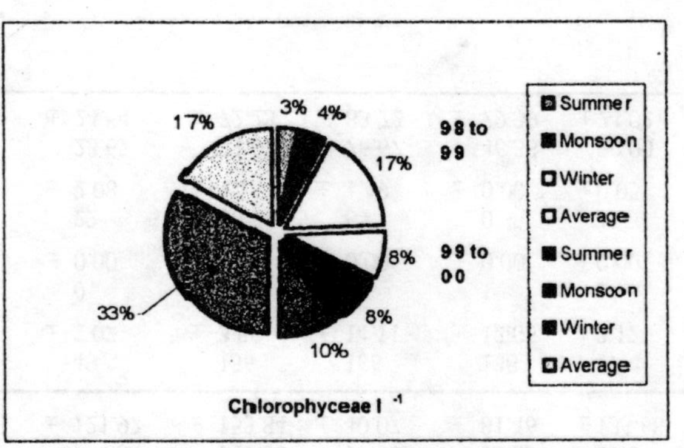

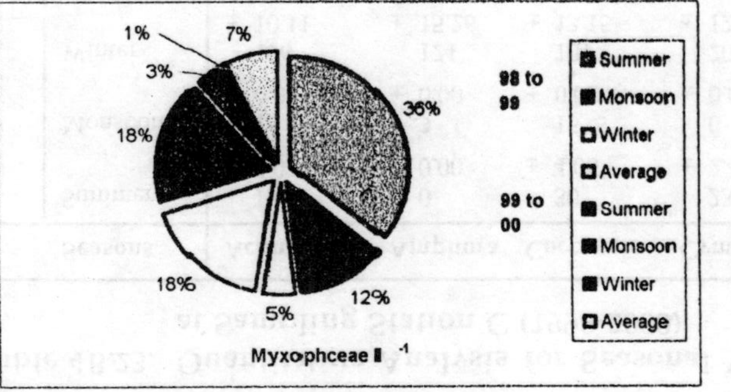

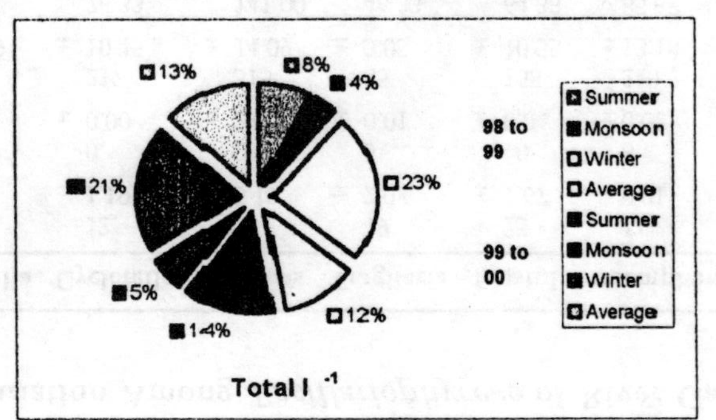

Fig 4B.28. PERCENTAGE COMPOSITION AMONG DIFFERENT GROUP IN PHYTOPLANKTON OF RIVER GANGA AT SAMPLING STATION C (1998-2000)

Table 4B.23. Quantitative Analysis for Seasonal Variation Among *Bacillariophyceae* of River Ganga at Sampling Station C (1998-2000)

	Seasons	Achnanthes	Amphora	Cocconeis	Cymbella	Cyclotella	Diatoms	Fragilaria	Frustulia	Gomphonena
1998 to 1999	Summer	15 ± 0.02	0 ± 0.00	30 ± 4.06	23 ± 2.44	12 ± 1.19	85 ± 5.19	49 ± 2.08	25 ± 1.67	11 ± 0.01
	Monsoon	16 ± 0.06	3 ± 0.00	15 ± 0.02	0 ± 0.00	0 ± 0.00	23 ± 2.62	5 ± 0.01	10 ± 0.04	6 ± 0.02
	Winter	196 ± 10.11	124 ± 15.26	104 ± 13.15	204 ± 12.19	217 ± 10.45	315 ± 14.09	85 ± 3.05	158 ± 10.55	240 ± 13.18
	Average I $^{-1}$	75.67 ± 104.21	42.33 ± 70.74	49.97 ± 47.65	75.67 ± 111.73	76.33 ± 121.97	141.00 ± 153.84	46.33 ± 40.07	64.33 ± 81.46	85.67 ± 133.68
1999 to 2000	Summer	104 ± 10.03	98 ± 5.66	76 ± 3.09	46 ± 1.06	49 ± 2.07	156 ± 3.58	166 ± 14.11	138 ± 12.05	47 ± 3.12
	Monsoon	0 ± 0.00	0 ± 0.00	1 ± 0.00	0 ± 0.00	0 ± 0.00	18 ± 0.02	0 ± 0.00	1 ± 0.00	1 ± 0.00
	Winter	86 ± 2.05	56 ± 1.06	117 ± 10.58	56 ± 12.06	22 ± 2.08	50 ± 1.16	64 ± 1.49	0 ± 0.00	12 ± 0.05
	Average I $^{-1}$	63.33 ± 55.58	51.33 ± 49.17	64.67 ± 58.82	34.00 ± 29.87	23.67 ± 24.54	74.67 ± 72.23	76.67 ± 83.72	46.33 ± 79.39	20.00 ± 24.02

Table 4B.23. (Contd.)

	Seasons	Navicula	Nitzschia	Pinnularia	Phacus	Stauroneis	Surirella	Synedra	Tabellaria	Total I $^{-1}$
1998 to 1999	Summer	56 ± 3.66	2 ± 0.01	42 ± 3.78	14 ± 0.05	59 ± 0.08	87 ± 2.09	20 ± 1.06	57 ± 3.19	587 ± 27.04
	Monsoon	0 ± 0.00	4 ± 0.00	6 ± 0.01	9 ± 0.04	4 ± 0.01	0 ± 0.00	8 ± 0.05	0 ± 0.00	109 ± 6.59
	Winter	193 ± 5.66	115 ± 8.59	137 ± 10.06	208 ± 12.04	115 ± 10.11	85 ± 9.49	166 ± 11.28	204 ± 15.37	2866 ± 61.89
	Average I $^{-1}$	83.00 ± 99.29	40.33 ± 64.67	61.67 ± 67.68	77.00 ± 113.48	59.33 ± 55.50	57.33 ± 49.66	64.67 ± 87.96	87.00 ± 105.26	1187.33 ± 23.61
1999 to 2000	Summer	167 ± 5.19	86 ± 3.48	37 ± 1.09	58 ± 2.69	56 ± 1.08	95 ± 1.02	107 ± 9.22	78 ± 4.37	1564 ± 43.07
	Monsoon	0 ± 0.00	2 ± 0.00	0 ± 0.00	0 ± 0.00	1 ± 0.00	1 ± 0.00	0 ± 0.00	1 ± 0.00	26 ± 4.29
	Winter	56 ± 0.05	24 ± 0.02	49 ± 0.01	81 ± 0.08	42 ± 0.05	52 ± 1.06	36 ± 1.01	56 ± 0.71	859 ± 28.23
	Average I $^{-1}$	74.33 ± 85.00	37.33 ± 43.56	28.67 ± 25.54	46.33 ± 41.74	33.00 ± 28.58	49.33 ± 47.06	47.67 ± 54.45	45.00 ± 39.66	816.33 ± 17.77

± = Standard Deviation.

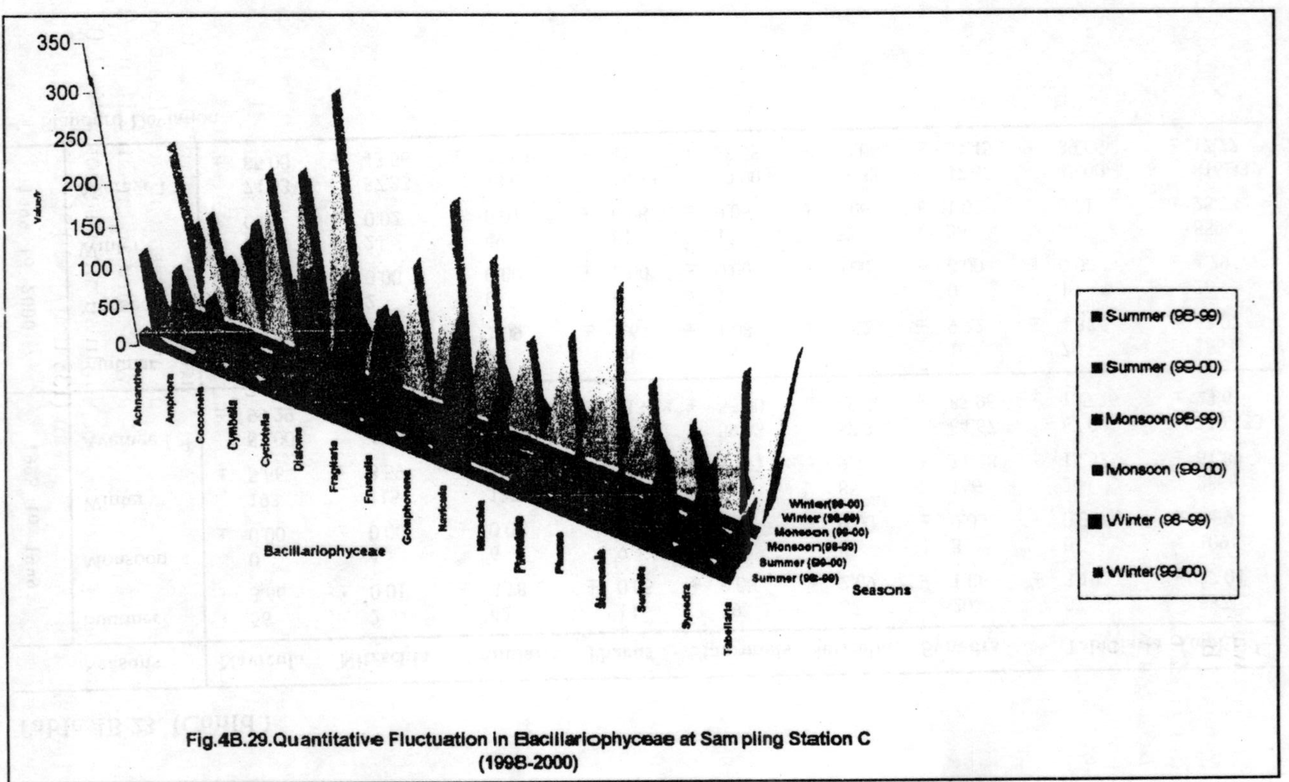

Fig.4B.29. Quantitative Fluctuation in Bacillariophyceae at Sampling Station C (1998-2000)

Surirella was recorded highest (87 unit/l ±2.09) and *Amphora* was found to be Nil; in monsoon season *Diatoms* was maximum (23 unit/l ±2.62) and *Cymbella, Cyclotella, Navicula, Surirella* and *Tabellaria* were recorded to be Nil. In winter season *Diatoms* noted as optimum concentration (315 unit/l ±14.09) and least concentration was noted for *Fragilaria* and *Surirella* (85 unit/l ±3.05 and 85 unit/l ±9.49). In 1999-2000 summer, *Fragilaria* were recorded highest (166 unit/l ±14.11) and *Pinnularia* was recorded lowest (37 unit/l ±1.09); in monsoon season *Diatoms* was noted maximum (18 unit/l ±0.02) while *Achnanthes, Amphora, Cymbella, Cyclotella, Fragilaria, Navicula, Pinnularia, Phacus* and *Synedra* were recorded Nil. In winter season *Cocconeis* was recorded in optimum concentration (117 unit/l ±10.58) and *Frustulia* was observed least as Nil.

The *Zygnema* observed to be highest average value (95.67 unit/l ±115.52) among **Chlorophyceae** in 1998-1999 followed by *Schizogonium* (88.67 unit/l ±70.15), *Cladophora* (86 unit/l ±86.85), *Spirogyra* (85.67 unit/l ±18.72), *Ulothrix* (83.33 unit/l ±30.09), *Vaucheria* (81 unit/l ±77.58), *Chlorella* (69.33 unit/l ±95.00), *Cosmarium* (61 unit/l ±92.97), *Closterium* (60.67 unit/l ±83.58), *Chaetophora* (52.33 unit/l ±70.12) and *Oedogonium* (5 unit/l ±8.66). In 1999-2000, it revealed *Spirogyra* was maximum (179.33 unit/l ±91.90) followed by *Ulothrix* (167.33 unit/l ±78.24), *Schizogonium* (197.33 unit/l ±116.07), *Cosmarium* (166.67 unit/l ±141.77), *Vaucheria* (161.33 unit/l ±115.78), *Closterium* (151.33 unit/l ±134.40), *Oedogonium* (144.67 unit/l ±148.60), *Chlorella* (129 unit/l ±133.07), *Chaetophora* (125.67 unit/l ±189.80), *Cladophora* (99.33 unit/l ±158.40) and *Zygnema* (95 unit/l ±55.34).

A great variation during different season was observed in **Chlorophyceae** at Muneshwar Ghat. In summer season 1998-1999, *Schizogonium* was recorded highest (94 unit/l ±10.56) and *Chaetophora, Oedogonium* and *Zygnema* were found lowest as Nil. In monsoon season *Cladophora,* was noted maximum (77 unit/l ±5.48) and *Closterium* and *Cosmarium* were found as Nil. In winter season *Zygnema* was noted optimum 224 unit/l ±15.61 and *Oedogonium* was Nil. In summer season 1999-2000, *Ulothrix* was noted maximum (138 unit/l ±10.38) and *Cladophora* was minimum (16 unit/l ±2.00). In monsoon season *Zygnema* was highest (142 unit/l ±15.12) and *Chaetophora* and *Cladophora* were found as Nil. In winter season *Chaetophora* was noted optimum 344 unit/l ±15.20 and *Zygnema* was observed least (109 unit/l ±10.37). The total highest reading for **Chlorophyceae** was observed in winter 1584 unit/l ±57.45, and lowest in summer season (306 unit/l ±35.09) for first year (1998-1999). In second year (1999-2000) maximum **Chlorophyceae** was detected during winter 3087 unit/l ±61.95 and minimum in summer 718 unit/l ±45.22 (Table 4B.24, Fig. 4B.30).

In **Myxophyceae,** revealing average value, *Anacystis* was noted optimum (100.67 unit/l ±107.55) in first year (1998-1999), followed by in descending order *Anabaena* (75.67 unit/l ±64.08), *Lyngbya* (69.67 unit/l ±53.54) and *Oscillatoria* (55 unit/l ±58.28). In second year (1999-2000) *Oscillatoria* was recorded highest (38 unit/l ±47.62), followed by *Anacystis* (31.67 unit/l ±41.02), *Anabaena* (30.33 unit/l ±40.25) and *Lyngbya* (28 unit/l ±32.74).

For seasonal fluctuation of **Myxophyceae** it is revealed that in summer of 1998-1999, *Anacystis* was existing in maximum (223 unit/l ±15.02) and *Oscillatoria* and *Lyngbya* were minimum (122 unit/l ±12.44 and 122 unit/l ±15.26). In monsoon season *Lyngbya* was observed to be highest (72 unit/l ±6.22), while *Oscillatoria* was lowest (27 unit/l ±2.06). In winter season *Anabaena* was optimum (26 unit/l ±2.22) and *Lyngbya* was found to be least (15 unit/l ±2.04). In second season of this study at Muneshwar Ghat it was recorded that in summer season *Oscillatoria* was maximum (92 unit/l ±11.02) and *Lyngbya* was minimum (64 unit/l ±3.09), in monsoon season *Oscillatoria* was highest (20 unit/l ±2.06) and *Lyngbya* was lowest as Nil, and in winter season *Lyngbya* was optimum (20 unit/l ±1.63) and least was *Anacystis* and *Anabaena* founds as Nil. In total value of 1998-2000 study the maximum **Myxophyceae** for 1998-1999 were observed in summer season (615 unit/l ±47.77) and minimum in winter season (78 unit/l ±5.07), and in 1999-2000 highest recorded was 310 unit/l ±11.47 in summer and lowest was 22 unit/l ±9.71 in winter season (Table 4B.25, Fig 4B.31).

(b) Microbiological

During analysis of Most Probable Number (MPN) and Standard Plate Count (SPC) at this sampling station observation revealed that mean MPN and SPC was higher (98.33 MNP/100 ml

Table 4B.24. Quantitative Analysis for Seasonal Variation Among *Chlorophyceae* of River Ganga at Sampling Station C (1998-2000)

	Seasons	Closterium	Cosmarium	Chaetophora	Chlorella	Cladophora	Oedogonium
1998 to 1999	Summer	26 ± 5.36	15 ± 2.05	0 ± 0.00	2 ± 1.00	4 ± 1.05	0 ± 0.00
	Monsoon	0 ± 0.00	0 ± 0.00	25 ± 3.89	28 ± 3.10	77 ± 5.48	15 ± 2.77
	Winter	156 ± 10.22	168 ± 12.46	132 ± 14.92	178 ± 15.12	177 ± 14.28	0 ± 0.00
	Average I $^{-1}$	60.67 ± 83.58	61.00 ± 92.97	52.33 ± 70.12	69.33 ± 95.00	86.00 ± 86.85	5.00 ± 8.66
1999 to 2000	Summer	85 ± 4.25	62 ± 2.82	33 ± 1.05	22 ± 1.59	16 ± 2.00	25 ± 2.31
	Monsoon	63 ± 3.33	110 ± 10.22	0 ± 0.00	87 ± 4.05	0 ± 0.00	98 ± 2.44
	Winter	306 ± 14.77	328 ± 14.61	344 ± 15.20	278 ± 12.34	282 ± 12.01	311 ± 15.28
	Average I $^{-1}$	151.33 ± 134.40	166.67 ± 141.77	125.67 ± 189.80	129.00 ± 133.07	99.33 ± 158.40	144.67 ± 148.60

Table 4B.24. (Contd.)

	Seasons	Schizogonium	Spirogyra	Ulothrix	Vaucheria	Zygnema	Total l $^{-1}$
1998 to 1999	Summer	94	78	68	19	0	306
		± 10.56	± 11.42	± 6.19	± 1.09	± 0.00	± 35.09
	Monsoon	16	72	64	56	63	416
		± 3.05	± 4.11	± 6.05	± 4.17	± 2.27	± 29.11
	Winter	156	107	118	168	224	1584
		± 11.06	± 9.32	± 9.88	± 14.28	± 15.61	± 57.45
	Average l $^{-1}$	88.67	85.67	83.33	81.00	95.67	768.67
		± 70.15	± 18.72	± 30.09	± 77.58	± 115.52	± 25.52
1999 to 2000	Summer	109	135	138	59	34	718
		± 10.48	± 1.05	± 10.38	± 5.16	± 2.11	± 45.22
	Monsoon	92	118	108	138	142	956
		± 5.79	± 10.44	± 10.28	± 12.44	± 15.12	± 48.39
	Winter	301	285	256	287	109	3087
		± 13.62	± 10.28	± 11.24	± 5.91	± 10.37	± 61.95
	Average l $^{-1}$	167.33	179.33	167.33	161.33	95.00	1587.00
		± 116.07	± 91.90	± 78.24	± 115.78	± 55.34	± 28.57

± = Standard Deviation.

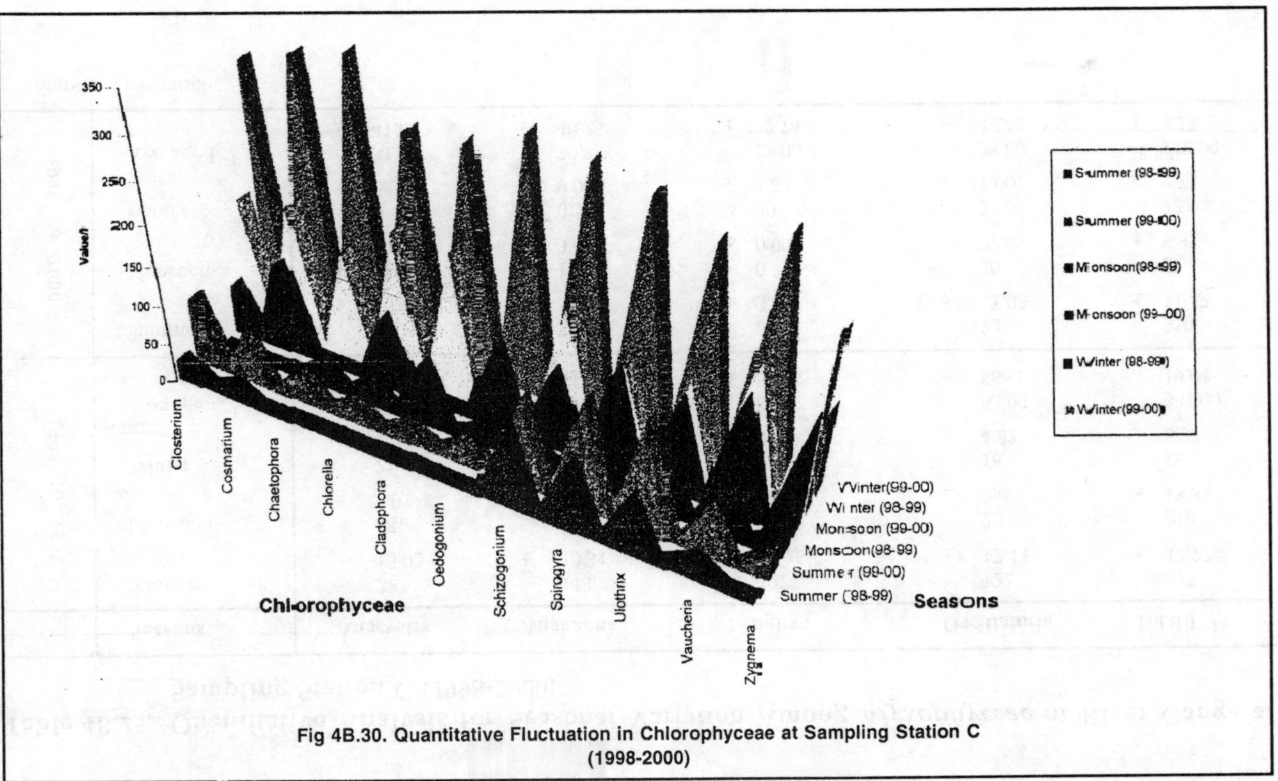

Fig 4B.30. Quantitative Fluctuation in Chlorophyceae at Sampling Station C (1998-2000)

Table 4B.25. Quantitative Analysis for Seasonal Variation Among *Myxophyceae* of River Ganga at Sampling Station C (1998-2000)

	Seasons	Anacystis	Anabaena	Lyngbya	Oscillatoria	Total l^{-1}
1998 to 1999	Summer	223 ± 15.02	148 ± 10.34	122 ± 15.26	122 ± 12.44	615 ± 47.77
	Monsoon	58 ± 2.02	53 ± 4.91	72 ± 6.22	27 ± 2.06	210 ± 18.81
	Winter	21 ± 1.03	26 ± 2.22	15 ± 2.04	16 ± 4.82	78 ± 5.07
	Average I^{-1}	100.67 ± 107.55	75.67 ± 64.08	69.67 ± 53.54	55.00 ± 58.28	301.00 ± 19.04
1999 to 2000	Summer	78 ± 10.03	76 ± 5.66	64 ± 3.09	92 ± 11.02	310 ± 11.47
	Monsoon	17 ± 2.00	15 ± 3.06	0 ± 0.00	20 ± 2.06	52 ± 8.91
	Winter	0 ± 0.00	0 ± 0.00	20 ± 1.63	2 ± 1.00	22 ± 9.71
	Average I^{-1}	31.67 ± 41.02	30.33 ± 40.25	28.00 ± 32.74	38.00 ± 47.62	128.00 ± 4.28

± = Standard Deviation.

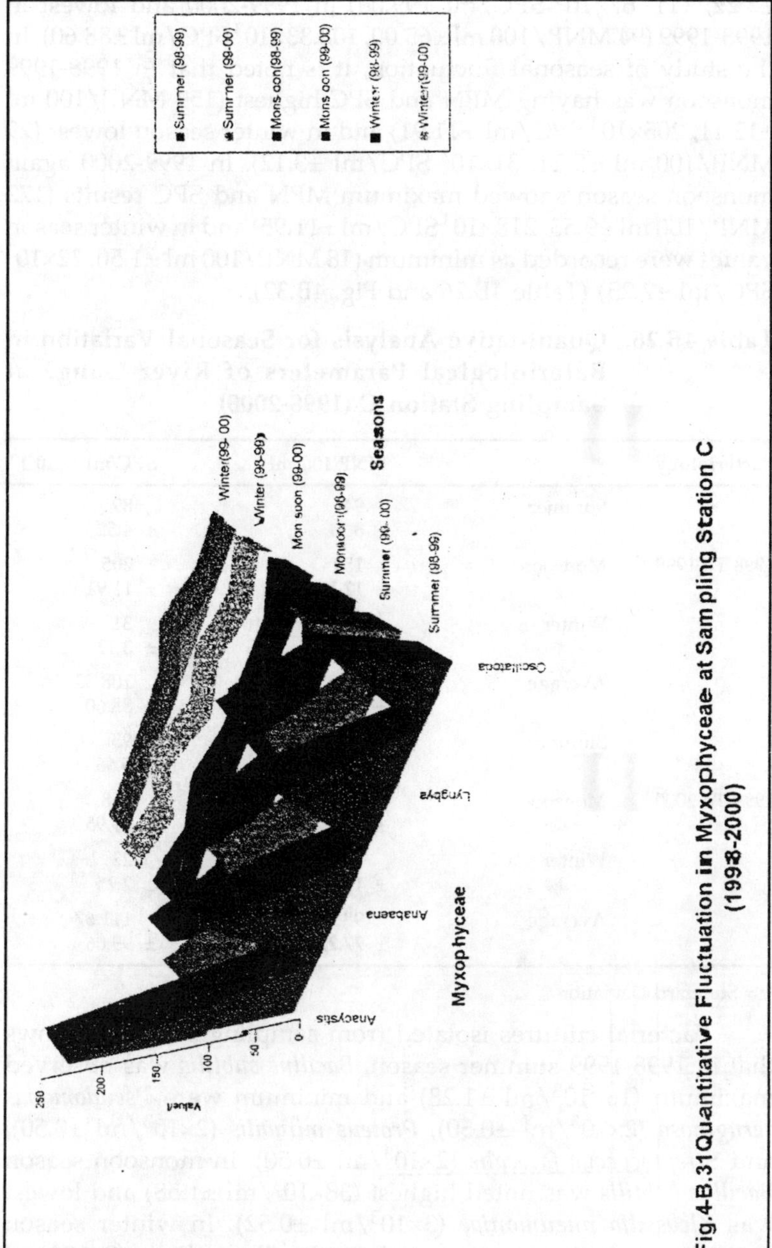

Fig.4-B.31 Quantitative Fluctuation in Myxophyceae at Sampling Station C (1998-2000)

±7.22, 111. 67×10^3 SPC/ml ±99.06) in 1999-2000 and lowest in 1998-1999 (94 MNP/100 ml ±65.00, 108.33×10^3 SPC/ml ±88.60). In the study of seasonal fluctuation, it is noted that in 1998-1999 monsoon was having MPN and SPC highest (159 MNP/100 ml ±12.11, 205×10^3 SPC/ml ±11.91) and in winter season lowest (29 MNP/100 ml ±5.24, 31×10^3 SPC/ml ±3.12). In 1999-2000 again monsoon season showed maximum MPN and SPC results (172 MNP/100 ml ±9.53, 218×10^3 SPC/ml ±11.95) and in winter season values were recorded as minimum (18 MNP/100 ml ±1.50, 22×10^3 SPC/ml ±2.25) (Table 4B.26 and Fig. 4B.32).

Table 4B.26. Quantitative Analysis for Seasonal Variation in Bateriological Parameters of River Ganga at Sampling Station C (1998-2000)

Bacteriology		MNP/100 ml	SPC/ml (×10^3)
	Summer	94 ± 8.24	89 ± 4.52
1998 To 1999	Monsoon	159 ± 12.11	205 ± 11.91
	Winter	29 ± 5.24	31 ± 3.12
	Average	94.00 ± 65.00	108.33 ± 88.60
	Summer	105 ± 11.26	95 ± 5.66
1999 To 2000	Monsoon	172 ± 9.53	218 ± 11.95
	Winter	18 ± 1.50	22 ± 2.25
	Average	98.33 ± 77.22	111.67 ± 99.06

± = Standard Deviation.

Bacterial cultures isolated from sampling station C shows that in 1998-1999 summer season, *Bacillus subtilis* was observed maximum (16×10^3/ml ±1.28) and minimum were *Pseudomonas aeruginosa* (2×10^3/ml ±0.50), *Proteus mirabilis* (2×10^3/ml ±0.50), and *Streptococcus facecalis* (2×10^3/ml ±0.50). In monsoon season *Bacillus subtilis* was noted highest (38×10^3/ml ±1.58) and lowest was *Klebssilla pneumoniae* (3×10^3/ml ±0.52). In winter season optimum growth was observed for *Bacillus subtilis* (9×10^3/ml

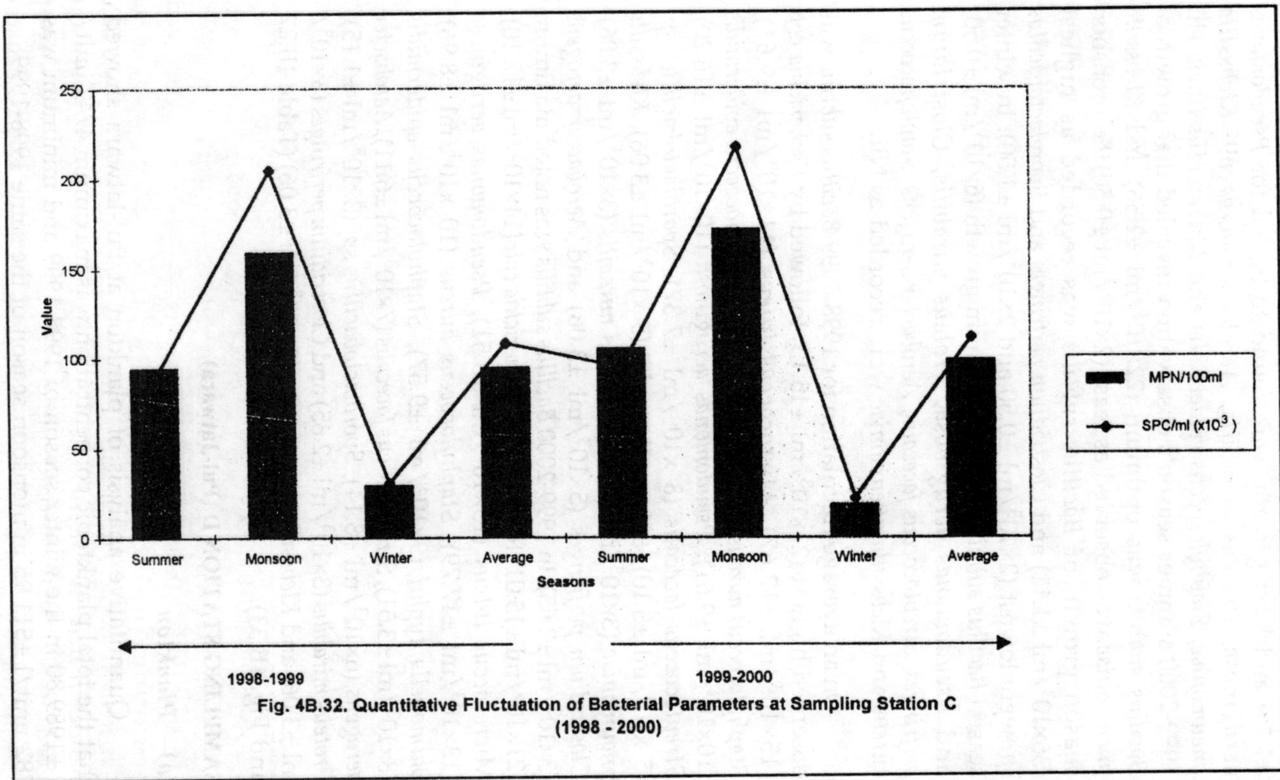

Fig. 4B.32. Quantitative Fluctuation of Bacterial Parameters at Sampling Station C (1998 - 2000)

±1.20) and least growth were found to be Nil for *Pseudomonas aeruginosa, Proteus mirabilis, Aerobacter aerogens, Klebssilla pneumoniae, Staphylococcus epidermidis* and *Micrococcus luteus*. In 1999-2000 summer season's observation revealed that growth of *Bacillus subtilis* was optimum (22×10^3/ml ±2.56) and *Klebssilla pneumoniae* was observed as least (2×10^3/ml ±0.50). In monsoon season growth of *Bacillus subtilis* was recorded as highest (36×10^3/ml ±1.11) and *Clostridium perfringes* and *Sporolactobacillus sp*. were lowest (2×10^3/ml ±0.50 and 2×10^3/ml ±1.50). In winter season *Bacillus subtilis* showed optimum growth (6×10^3/ml ±0.50) and *Pseudomonas aeruginosa, Proteus mirabilis, Clostridium perfringes, Streptococcus facecalis, Aerobacter aerogens, Staphylococcus aureus* and *Klebssilla pneumoniae* were recorded as Nil.

On an average calculation for 1998-1999 *Bacillus subtilis* was observed highest (21×10^3/ml ±15.13) followed by *Escherichia coli* (15×10^3/ml ±17.67), *Micrococcus luteus* (11×10^3/ml ±9.61), *Staphylococcus aureus* (10×10^3/ml ±2.08), *Staphylococcus epidermidis* (10×10^3/ml ±9.61), *Pseudomonas aeruginosa* (10×10^3/ml ±16.20), *Streptococcus facecalis* (8×10^3/ml ±7.37), *Sporolactobacillus sp.* (5×10^3/ml ±3.10), *Salmonella typhii* (5×10^3/ml ±3.06), *Klebssilla pneumoniae* (3×10^3/ml ±2.52), *Proteus mirabilis* (3×10^3/ml ±3.06), *Clostridium perfringes* (5×10^3/ml ±3.06) and *Aerobacter aerogens* (3×10^3/ml ±2.65). In 1999-2000 *Bacillus subtilis* was noted maximum (21×10^3/ml ±15.01) followed by *Escherichia coli* (14×10^3/ml ±12.10), *Micrococcus luteus* (13×10^3/ml ±3.51), *Pseudomonas aeruginosa* (13×10^3/ml ±17.79), *Staphylococcus aureus* (10×10^3/ml ±8.96), *Salmonella typhii* (9×10^3/ml ±9.57), *Staphylococcus epidermidis* (8×10^3/ml ±3.51), *Streptococcus facecalis* (7×10^3/ml ±6.11), *Aerobacter aerogens* (6×10^3/ml ±8.14), *Sporolactobacillus sp.* (3×10^3/ml ±1.15), *Proteus mirabilis* (3×10^3/ml ±2.65) and *Clostridium perfringes* (3×10^3/ml ±3.06) and *Klebssilla pneumoniae* (3×10^3/ml ±3.06) (Table 4B.27 and Fig. 4B.33).

SAMPLING STATION D (Pul-Jatwara)

(a) Plankton

Quantitative analysis of plankton at Pul-Jatwara showed that the total planktonic concentration was maximum 4742 unit/l ±1989.80 in the winter season of 1998-1999 and minimum was 782 unit/l ±511.95 in monsoon season of the same 1998-1999.

Table 4B.27. Isolation of Bacterial Culture from Sampling Station C (1998-2000)

Bacteria	1998 to 1999				1999 to 2000			
	Summer	Monsoon	Winter	Average	Summer	Monsoon	Winter	Average
Escherichia coli*	6 ± 1.50	35 ± 2.53	3 ± 0.50	15 ± 17.67	10 ± 1.50	28 ± 2.22	5 ± 1.00	14 ± 12.10
Pseudomonas aeruginosa*	2 ± 0.50	29 ± 2.51	0 ± 0.00	10 ± 16.20	5 ± 1.00	33 ± 1.52	0 ± 0.00	13 ± 17.79
Bacillus subtilis**	16 ± 1.28	38 ± 1.58	9 ± 1.20	21 ± 15.13	22 ± 2.56	36 ± 1.11	6 ± 0.50	21 ± 15.01
Proteus mirabilis**	2 ± 0.50	6 ± 0.56	0 ± 0.00	3 ± 3.06	4 ± 0.50	5 ± 0.50	0 ± 0.00	3 ± 2.65
Clostridium perfringes*	4 ± 0.52	8 ± 1.00	2 ± 0.52	5 ± 3.06	6 ± 0.50	2 ± 0.50	0 ± 0.00	3 ± 3.06
Streptococcus facecalis*	2 ± 0.50	16 ± 2.11	5 ± 0.51	8 ± 7.37	8 ± 1.50	12 ± 1.50	0 ± 0.00	7 ± 6.11
Aerobacter aerogens*	4 ± 0.50	5 ± 0.53	0 ± 0.00	3 ± 2.65	2 ± 0.50	15 ± 1.50	0 ± 0.00	6 ± 8.14
Salmonella typhii*	6 ± 0.52	8 ± 1.20	2 ± 1.00	5 ± 3.06	4 ± 1.00	20 ± 1.28	2 ± 0.50	9 ± 9.87
Staphylococcus aureus*	9 ± 0.54	12 ± 0.56	8 ± 1.00	10 ± 2.08	15 ± 1.50	16 ± 0.50	0 ± 0.00	10 ± 8.96
Sporolactobacillus sp.**	8 ± 1.50	5 ± 0.50	2 ± 1.00	5 ± 3.00	4 ± 0.50	2 ± 1.50	2 ± 0.50	3 ± 1.15
Klebssilla pneumoniae*	5 ± 0.50	3 ± 0.52	0 ± 0.00	3 ± 2.52	2 ± 0.50	6 ± 0.50	0 ± 0.00	3 ± 3.06
Staphylococcus epidermidis*	12 ± 1.55	19 ± 1.54	0 ± 0.00	10 ± 9.61	8 ± 1.50	12 ± 1.26	5 ± 0.50	8 ± 3.51
Micrococcus luteus**	13 ± 2.30	21 ± 1.50	0 ± 0.00	11 ± 9.61	5 ± 0.58	31 ± 1.50	2 ± 0.50	13 ± 3.51
Total ml $^{-1}$ ($\times 10^3$)	89 ± 4.52	205 ± 11.91	31 ± 3.12	108.33 ± 88.60	95 ± 5.66	218 ± 11.95	22 ± 2.25	111.67 ± 99.06

± = Standard Deviation
Unit for value is ml^{-1} ($\times 10^3$)

* = Pathogen
** = Non-Pathogen.

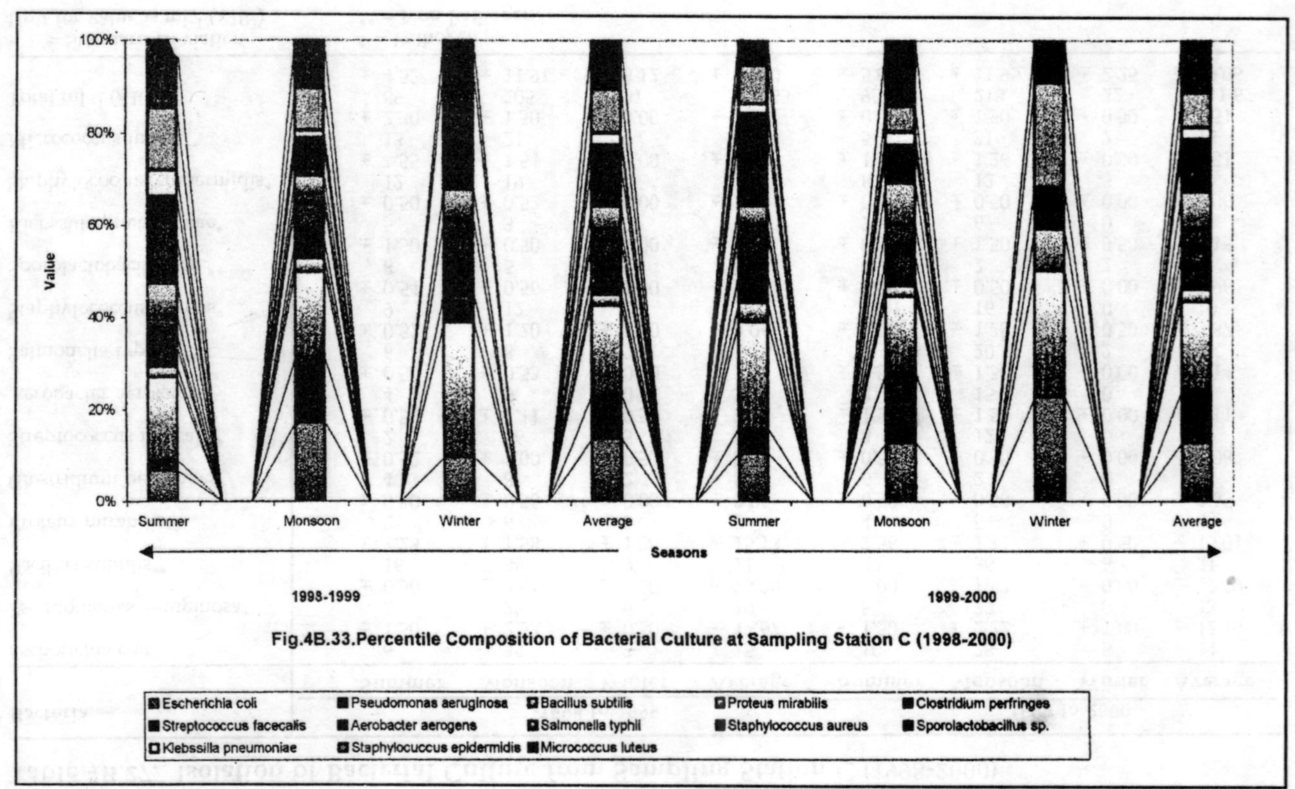

Fig.4B.33.Percentile Composition of Bacterial Culture at Sampling Station C (1998-2000)

Zooplankton was noted lowest 21 unit/l ±4.11 in the monsoon season of 1999-2000 and highest in the winter season of 1999-2000 (1134 unit/l ±121.52). In 1998-1999 phytoplankton was observed minimum in monsoon season (753 unit/l ±301.91) and maximum in 1998-1999 for the season of winter (3778 unit/l ±544.36) (Table 4B.28 and Figs. 4B.34, 4B.35).

1. Zooplankton

Among zooplankton phylum **Rotifera** was dominating and phylum **Copepoda** was least in both year. At this spot **Protozoa** was maximum in winter season of 1999-2000 (254 unit/l ±11.31) and minimum was found Nil in monsoon season (1998-1999). **Rotifera** was found to be highest in the winter season of 1999-2000 (453 unit/l ±33.78) and lowest was recorded in the monsoon season of 1998-1999 (7 unit/l ±2.52). **Cladocera** was observed in their optimum peak during 1999-2000 winter season (263 unit/l ±6.36) and in lowest peak (4 unit/l ±1.41) during monsoon season of the same year (1999-2000). **Copepoda** was maximum (164 unit/l ±11.14) in 1999-2000, winter season and minimum was found in monsoon season of 1998-1999 (7 unit/l ±2.11). In total at this sampling station winter season of 1999-2000 showed maximum (1134 unit/l ±121.52) and minimum in monsoon season of 1999-2000 (21 unit/l ±4.11) (Table 4B.29 and Fig. 4B.36).

The average value revealed that *Cyclops* was dominating in the zooplankton, 1998-1999 (66 unit/l ±77.86) followed by *Bosmina* (63 unit/l ±79.98), *Keratella* (60 unit/l ±82.04), *Amoeba* (47 unit/l ±68.25), *Daphnia* (41 unit/l ±53.86), *Philodima* (41 unit/l ±54.05), *Arcella* (36 unit/l ±47.06) and *Notholca* (34 unit/l ±42.90). In second year 1999-2000 *Keratella* (78 unit/l ±93.74) found in strength followed by *Philodima* (74 unit/l ±75.08), *Cyclops* (64 unit/l ±86.56), *Daphnia* (54 unit/l ±71.82), *Notholca* (48 unit/l ±59.81), *Arcella* (47 unit/l ±76.27), *Bosmina* (45 unit/l ±71.02) and *Amoeba* (44 unit/l ±65.04) (Table 4B.30 and Fig. 4B.37).

2. Phytoplankton

In phytoplankton **Chlorophyceae** was dominating and class **Bacillariophyceae** was found least in 1998-1999 and **Myxophyceae** was lowest in 1999-2000. At this sampling station

Table 4B.28. Seasonal Variation in Plankton for River Ganga at Sampling Station D (1998-2000)

Plankton	1998 to 1999				1999 to 2000			
	Summer	Monsoon	Winter	Average	Summer	Monsoon	Winter	Average
1. Zooplankton l^{-1}	166 ± 11.27	29 ± 6.13	964 ± 77.80	386.33 ± 504.94	209 ± 56.72	21 ± 4.11	1134 ± 121.52	454.67 ± 595.78
2. Phytoplankton l^{-1}	1299 ± 454.91	753 ± 301.91	3778 ± 544.36	1943.33 ± 1612.15	1610 ± 545.60	1715 ± 307.87	3293 ± 994.42	2206.00 ± 942.83
Total l^{-1}	1465 ± 801.15	782 ± 511.95	4742 ± 1989.80	2329.67 ± 2116.87	1819 ± 990.66	1736 ± 1197.84	4427 ± 1526.64	2660.67 ± 1530.25

± = Standard Deviation.

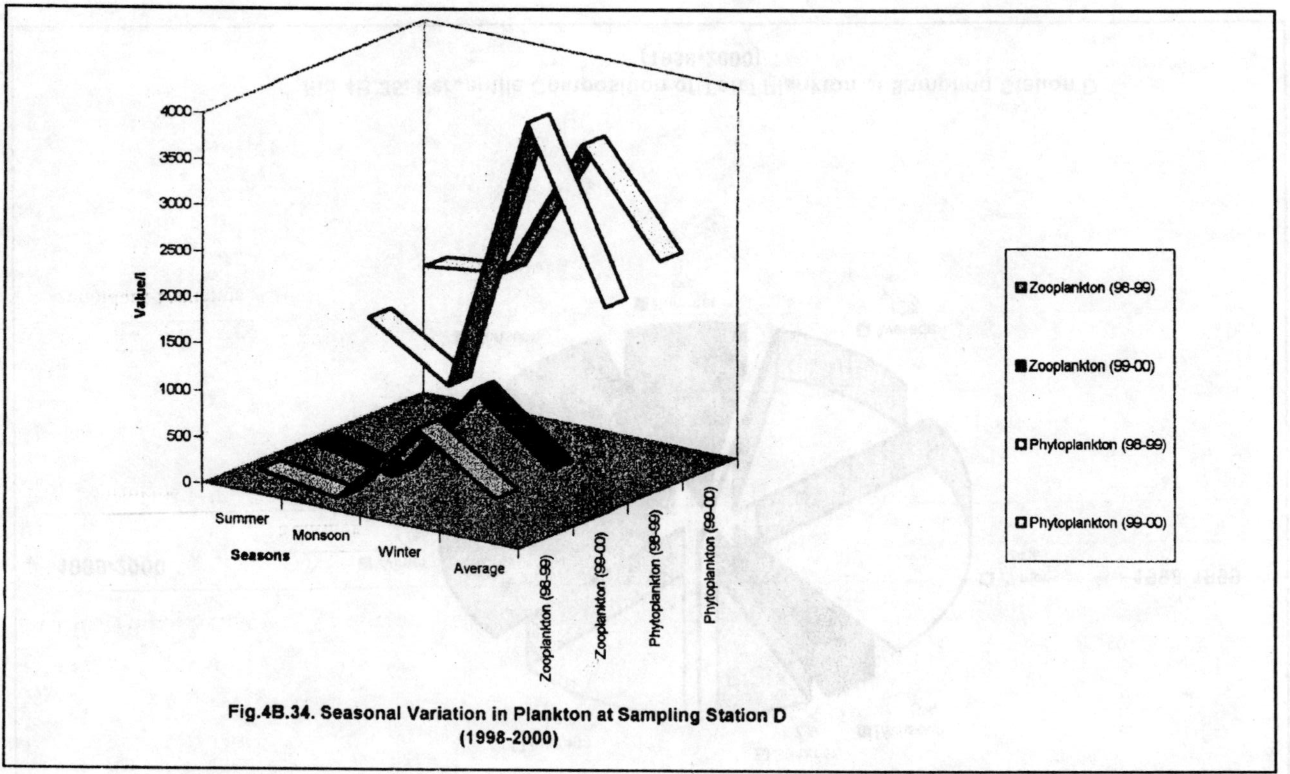

Fig.4B.34. Seasonal Variation in Plankton at Sampling Station D (1998-2000)

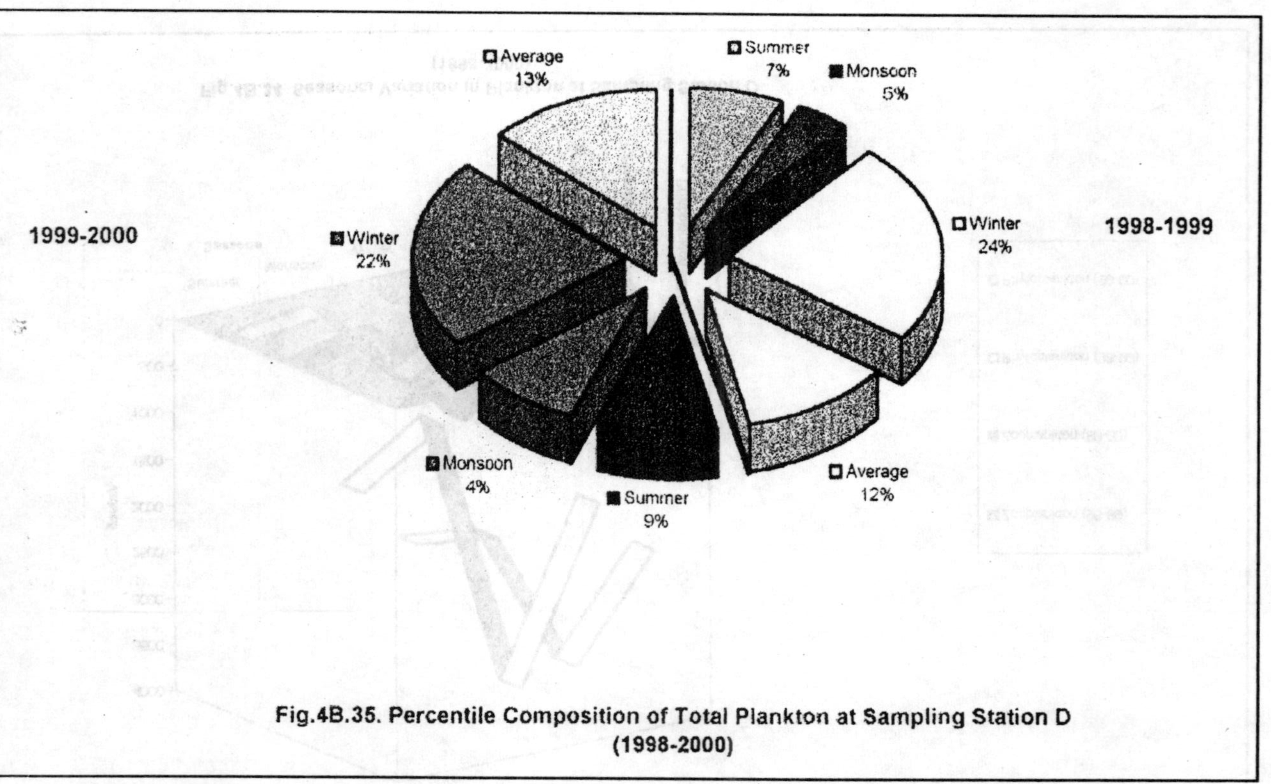

Fig.4B.35. Percentile Composition of Total Plankton at Sampling Station D (1998-2000)

Table 4B.29. Quantitative Analysis for Zooplankton in Different Seasons of River Ganga at Sampling Station D (1998-2000)

	Seasons	Protozoa	Rotifera	Cladocera	Copepoda	Total l $^{-1}$
1998 to 1999	Summer	33 ± 2.12	58 ± 1.15	39 ± 2.12	36 ± 1.50	166 ± 11.27
	Monsoon	0 ± 0.00	7 ± 2.52	15 ± 10.61	7 ± 2.11	29 ± 6.13
	Winter	214 ± 25.46	339 ± 36.76	257 ± 37.48	154 ± 9.28	964 ± 77.80
	Average l $^{-1}$	82.33 ± 115.21	134.67 ± 178.79	103.67 ± 15.32	65.67 ± 77.86	386.33 ± 29.77
1999 to 2000	Summer	20 ± 5.66	137 ± 14.57	31 ± 14.85	21 ± 1.62	209 ± 56.72
	Monsoon	0 ± 0.00	9 ± 5.20	4 ± 1.41	8 ± 1.50	21 ± 4.11
	Winter	254 ± 11.31	453 ± 33.78	263 ± 6.36	164 11.14	1134 ± 121.52
	Average l $^{-1}$	91.33 ± 141.23	199.67 ± 228.54	99.33 ± 142.38	64.33 ± 86.56	454.67 ± 59.26

± = Standard Deviation.

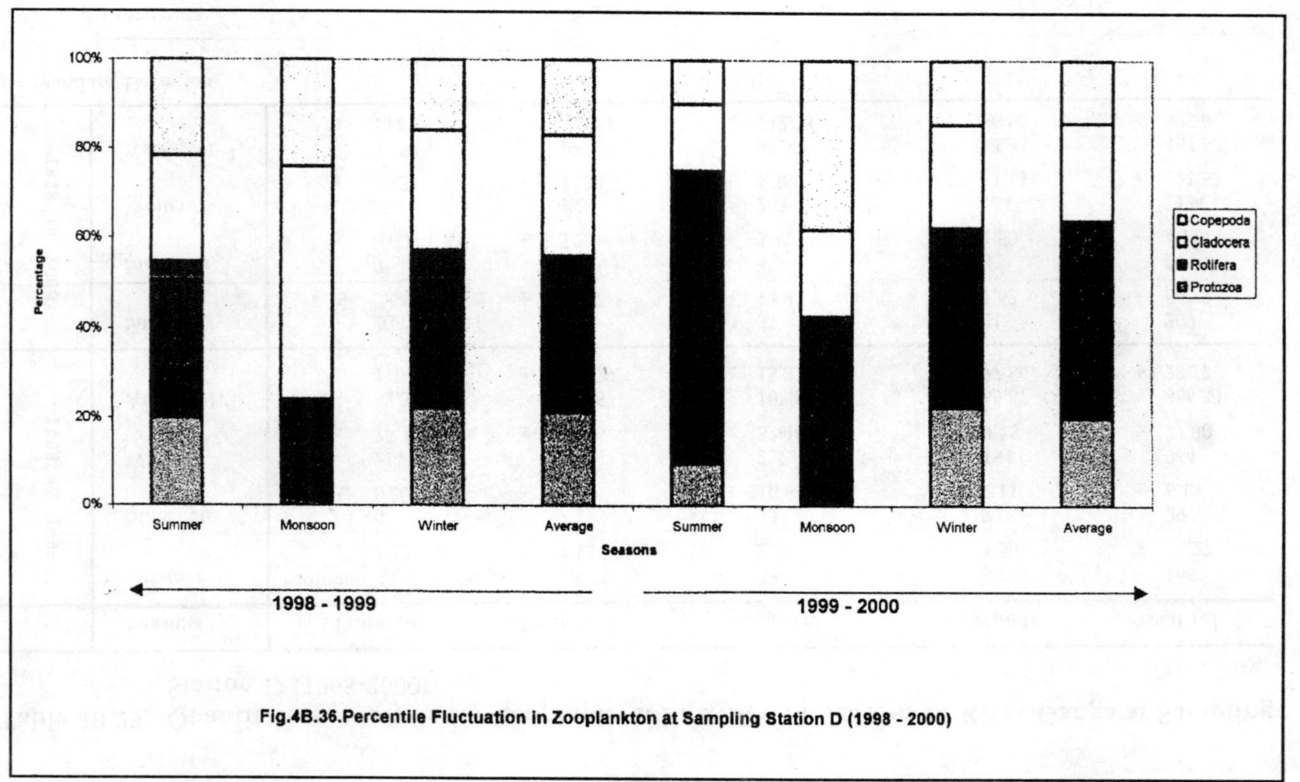

Fig.4B.36.Percentile Fluctuation in Zooplankton at Sampling Station D (1998 - 2000)

Table 4B.30. Quantitative Analysis for Zooplankton in Different Season of River Ganga at Sampling Station D (1998-2000)

		Protozoa		Rotifera			Cladocera		Copepoda	
	Seasons	Arcella	Amoeba	Keratella	Philodima	Notholca	Daphnia	Bosmina	Cyclops	Total I $^{-1}$
1998 to 1999	Summer	18 ± 3.24	15 ± 4.19	20 ± 2.09	20 ± 3.16	18 ± 1.08	21 ± 2.06	18 ± 1.06	36 ± 1.50	166 ± 6.43
	Monsoon	0 ± 0.00	0 ± 0.00	5 ± 0.50	0 ± 0.00	2 ± 1.00	0 ± 0.00	15 ± 1.50	7 ± 2.11	29 ± 5.32
	Winter	89 ± 7.01	125 ± 11.56	154 ± 12.51	102 ± 8.05	83 ± 5.12	102 ± 10.47	155 ± 11.29	154 ± 9.28	964 ± 30.56
	Average I $^{-1}$	36 ± 47.06	47 ± 68.25	60 ± 82.04	41 ± 54.05	34 ± 42.90	41 ± 53.86	63 ± 79.98	66 ± 77.86	386 ± 12.57
1999 to 2000	Summer	6 ± 1.50	14 ± 1.50	52 ± 4.55	56 ± 5.02	29 ± 3.14	26 ± 2.01	5 ± 1.00	21 ± 1.62	209 ± 19.25
	Monsoon	0 ± 0.00	0 ± 0.00	0 ± 0.00	9 ± 1.00	0 ± 0.00	1 ± 0.50	3 ± 1.00	8 ± 1.50	21 ± 3.78
	Winter	135 ± 11.25	119 ± 9.88	182 ± 7.56	156 ± 10.24	115 ± 10.58	136 ± 10.22	127 ± 9.27	164 ± 11.14	1134 ± 23.44
	Average I $^{-1}$	47 ± 76.27	44 ± 65.04	78 ± 93.74	74 ± 75.08	48 ± 59.81	54 ± 71.82	45 ± 71.02	64 ± 86.56	455 ± 13.44

± = Standard Deviation.

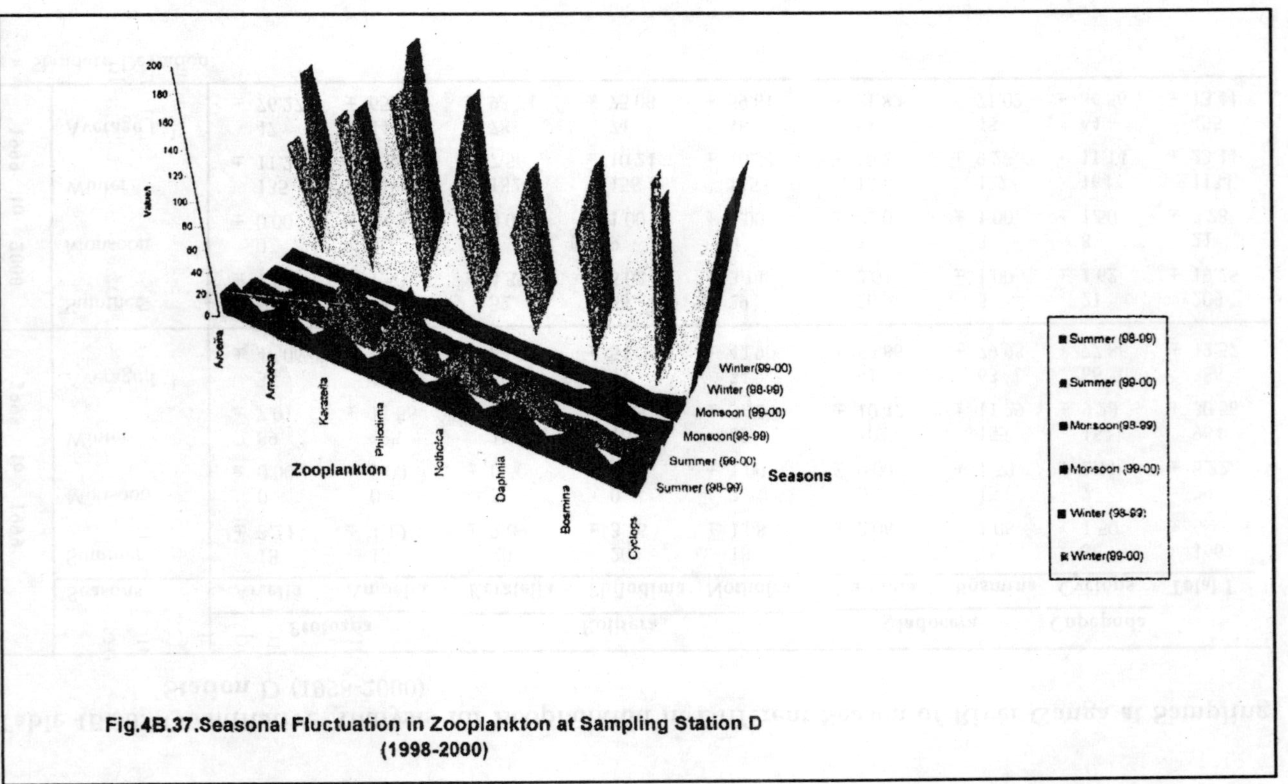

Fig.4B.37.Seasonal Fluctuation in Zooplankton at Sampling Station D (1998-2000)

Bacillariophyceae was observed maximum in winter season of 1999-2000 (928 unit/l ±27.91) and minimum in monsoon season (95 unit/l ±7.35) of 1998-1999. **Chlorophyceae** was found to be highest in winter season of 1999-2000 (2166 unit/l ±69.03) and lowest in monsoon season of 1998-1999 (599 unit/l ±40.56). **Myxophyceae** showed optimum peak during winter season of 1998-1999 (1029 unit/l ±53.04) and least peak during monsoon season of same year (59 unit/l ±13.65) (Table 4B.31 and Figs. 4B.38, 4B.39).

After observing mean values it was noted that *Diatoms* was dominated (67 unit/l ±71.88) among **Bacillariophyceae** in 1998-1999 followed by *Gomphonema* (28.33 unit/l ±39.93), *Stauroneis* (27 unit/l ±42.51), *Tabellaria* (25.33 unit/l ±37.00), *Pinnularia* (25.33 unit/l ±40.43), *Amphora* (21.33 unit/l ±23.18), *Navicula* (20.33 unit/l ±35.22), *Nitzschia* (20.67 unit/l ±22.74), *Phacus* (20.67 unit/l ±23.03), *Frustulia* (18 unit/l ±29.46), *Synedra* (17 unit/l ±9.54), *Fragilaria* (14.33 unit/l ±2.08), *Cymbella* (13 unit/l ±19.05), *Cyclotella* (12 unit/l ±12.49), *Surirella* (11.67 unit/l ±15.31), *Cocconeis* (11.33 unit/l ±11.02) and *Achnanthes* (11 unit/l ±12.9). In second year 1999-2000, *Diatoms* (59 unit/l ±47.00) were recorded as a dominating followed by *Frustulia* (46.67 unit/l ±31.63), *Cymbella* (43.67 unit/l ±38.99), *Surirella* (39 unit/l ±15.72), *Synedra* (38.67 unit/l ±30.09), *Stauroneis* (32.67 unit/l ±13.61), *Pinnularia* (32 unit/l ±55.43), *Cyclotella* (27.67 unit/l ±15.31), *Achnanthes* (27.33 unit/l ±28.02), *Tabellaria* (26.33 unit/l ±23.12), *Nitzschia* (25.67 unit/l ±35.92), *Phacus* (19.67 unit/l ±32.35), *Navicula* (19 unit/l ±28.69), *Cocconeis* (17 unit/l ±25.24), *Fragilaria* (15 unit/l ±11.79), *Amphora* (14.67 unit/l ±12.70) and *Gomphonema* (11.33 unit/l ±10.12) (Table 4B. 32 and Fig. 4B.40).

A great fluctuation among **Bacillariophyceae** was observed in seasonal study at this sampling point. In 1998-1999 summer, *Diatoms* was recorded highest (25 unit/l ±3.14) and *Frustulia, Navicula* were found to be Nil; in monsoon season *Diatoms* was maximum (26 unit/l ±1.04) and *Amphora, Cocconeis, Gomphonema, Navicula, Stauroneis* and *Surirella* were recorded to be Nil. In winter season *Diatoms* noted as optimum concentration (150 unit/l ±9.46) and least concentration was noted for *Fragilaria* (12 unit/l ±1.08). In 1999-2000 summer, *Diatoms* was recorded

Table 4B.31. Comparative Analysis for Seasonal Variation Among Phytoplankton of River Ganga at Sampling Station D (1998-2000)

	Seasons	Bacillariophyceae l^{-1}	Chlorophyceae l^{-1}	Myxophyceae l^{-1}	Total l^{-1}
1998 To 1999	Summer	130 ± 7.52	956 ± 38.97	213 ± 18.48	1299 ± 454.91
	Monsoon	95 ± 7.35	599 ± 40.56	59 ± 13.65	753 ± 301.91
	Winter	868 ± 32.40	1881 ± 62.49	1029 ± 53.04	3778 ± 544.36
	Average	364.33 ± 436.54	1145.33 ± 661.64	433.55 ± 521.37	1943 ± 432.32
1999 To 2000	Summer	327 ± 17.95	1156 ± 26.44	127 ± 16.46	1610 ± 545.60
	Monsoon	231 ± 14.80	654 ± 39.62	830 ± 23.07	1715 ± 307.87
	Winter	928 ± 27.91	2166 ± 69.03	199 ± 22.07	3293 ± 994.42
	Average	495.33 ± 377.76	1325.33 ± 770.09	385.33 ± 386.77	2206 ± 513.91

± = Standard Deviation.

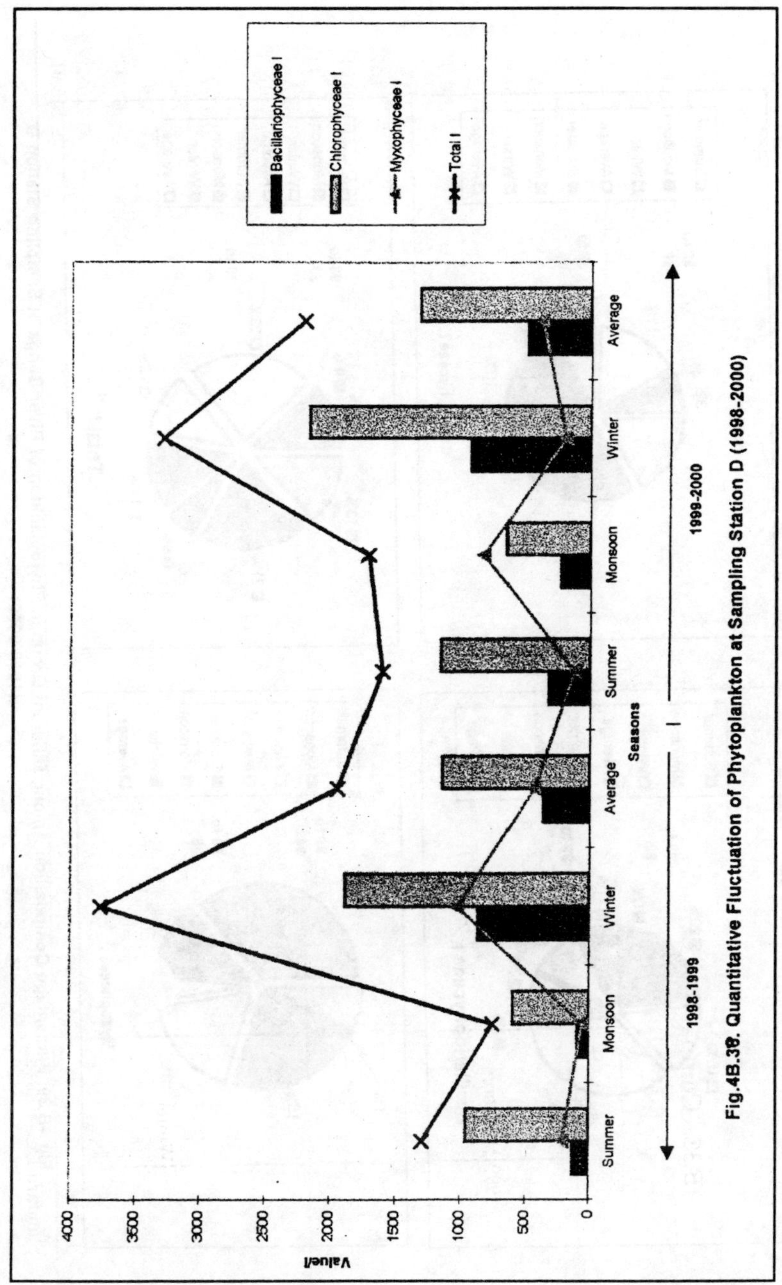

Fig.4B.38. Quantitative Fluctuation of Phytoplankton at Sampling Station D (1998-2000)

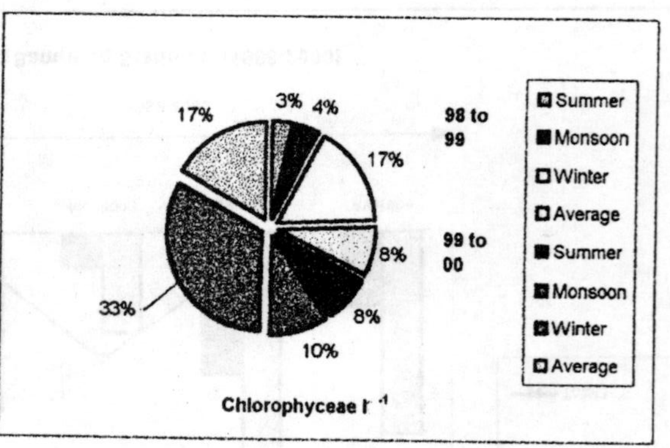

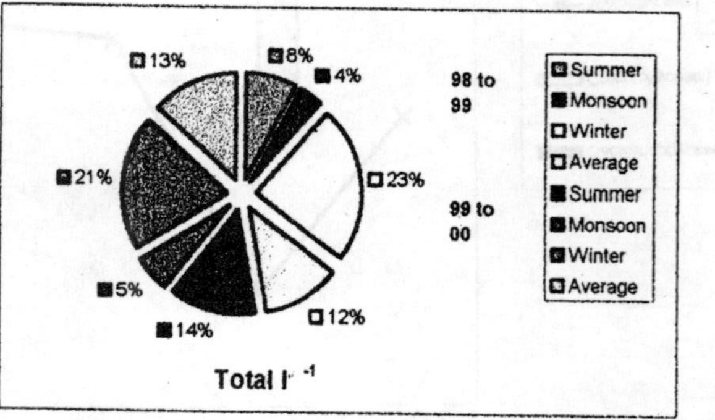

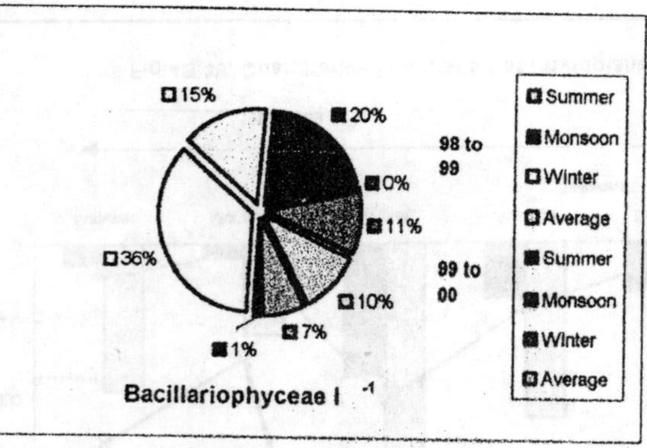

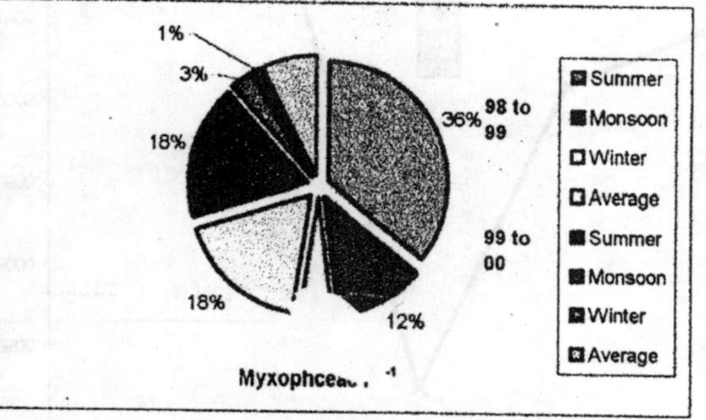

Fig 4B.39. Percentage Composition Among Different Group in Phytoplankton of River Ganga at Sampling Station D (1998-2000)

Table. 4B.32. Quantitative Analysis for Seasonal Variation Among *Bacillariophyceae* of River Ganga at Sampling Station D (1998-2000)

	Seasons	Achnanthes	Amphora	Cocconeis	Cymbella	Cyclotella	Diatoms	Fragilaria	Frustulia	Gomphonena
1998 to 1999	Summer	2 ± 0.02	18 ± 0.04	12 ± 0.01	2 ± 2.44	2 ± 1.12	25 ± 3.14	16 ± 0.02	0 ± 0.09	11 ±1.02
	Monsoon	6 ± 1.11	0 ± 0.00	0 ± 0.00	2 ± 0.01	8 ± 0.01	26 ± 1.04	15 ± 0.01	2 ± 0.04	0 ±0.00
	Winter	25 ± 2.05	46 ± 2.55	22 ± 1.61	35 ± 1.21	26 ± 0.06	150 ± 9.46	12 ± 1.08	52 ± 2.08	74 ±1.42
	Average l^{-1}	11.00 ± 12.29	21.33 ± 23.18	11.33 ± 11.02	13.00 ± 19.05	12.00 ± 12.49	67.00 ± 71.88	14.33 ± 2.08	18.00 ± 29.46	28.33 ±39.93
1999 to 2000	Summer	26 ± 5.23	22 ± 1.06	0 ± 0.00	56 ± 10.15	22 ± 0.04	59 ± 5.26	18 ± 1.14	21 ± 1.28	19 ±0.06
	Monsoon	0 ± 0.00	0 ± 0.00	5 ± 0.00	0 ± 0.00	45 ± 0.86	12 ± 0.07	2 ± 0.00	37 ± 3.06	15 ±4.16
	Winter	56 ± 14.16	22 ±13.11	46 ± 15.25	75 ±-14.19	16 ± 8.47	106 ± 12.11	25 ± 10.01	82 ± 12.12	0 ±13.14
	Average l^{-1}	27.33 ± 28.02	14.67 ± 12.70	17.00 ± 25.24	43.67 ± 38.99	27.67 ± 15.31	59.00 ± 47.00	15.00 ± 11.79	46.67 ± 31.63	11.33 ±10.02

Table 4B.32. (Contd.)

	Seasons	Navicula	Nitzschia	Pinnularia	Phacus	Stauroneis	Surirella	Synedra	Tabellaria	Total I $^{-1}$
1998 to 1999	Summer	0 ± 0.02	2 ± 0.00	1 ± 0.00	15 ± 0.01	5 ± 0.06	6 ± 0.01	11 ± 0.06	2 ± 0.01	130 ± 7.52
	Monsoon	0 ± 0.00	14 ± 0.03	3 ± 0.02	1 ± 0.00	0 ± 0.00	0 ± 0.00	12 ± 0.03	6 ± 0.02	95 ± 7.35
	Winter	61 ± 1.26	46 ± 0.02	72 ± 0.00	46 ± 2.19	76 ± 1.85	29 ± 1.22	28 ± 2.56	68 ± 3.48	868 ± 32.40
	Average I $^{-1}$	20.33 ± 35.22	20.67 ± 22.74	25.33 ± 40.43	20.67 ± 23.03	27.00 ± 42.51	11.67 ± 15.31	17.00 ± 9.54	25.33 ± 37.00	364.33 ± 13.08
1999 to 2000	Summer	0 ± 0.00	2 ± 0.00	0 ± 0.00	0 ± 0.00	28 ± 0.02	32 ± 0.05	10 ± 0.04	12 ± 0.53	327 ± 17.95
	Monsoon	5 ± 0.02	8 ± 0.01	0 ± 0.00	2 ± 0.00	22 ± 2.44	28 ± 5.16	36 ± 3.03	14 ± 0.06	231 ± 14.80
	Winter	52 ± 10.15	67 ± 6.06	96 ± 4.92	57 ± 13.15	48 ± 3.11	57 ± 14.56	70 ± 15.87	53 ± 14.62	928 ± 27.91
	Average I $^{-1}$	19.00 ± 28.69	25.67 ± 35.92	32.00 ± 55.43	19.67 ± 32.35	32.67 ± 13.61	39.00 ± 15.72	38.67 ± 30.09	26.33 ± 23.12	495.33 ± 13.01

± = Standard Deviation.

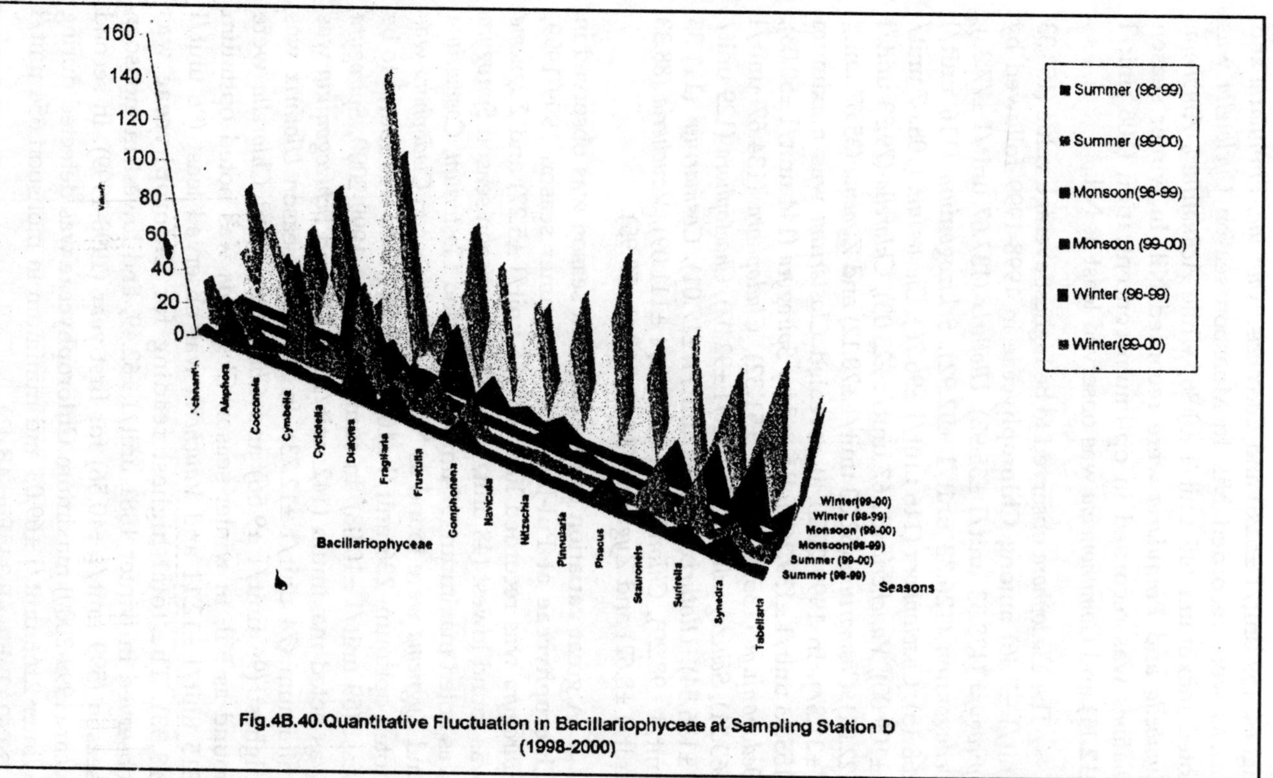

Fig.4B.40.Quantitative Fluctuation in Bacillariophyceae at Sampling Station D (1998-2000)

highest (59 unit/l ±5.26) and *Cocconeis, Navicula, Pinnularia* and *Phacus* were recorded Nil. In Monsoon season *Cyclotella* was noted maximum (45 unit/l ±0.86) while *Achnanthes, Amphora, Cymbella* and *Pinnularia* were recorded Nil. In winter season *Diatoms* was recorded in optimum concentration (106 unit/l ±12.11) and *Gomphonema* was observed least as Nil.

The *Cladophora* observed to be highest average value (153.33 unit/l ±7.36) among **Chlorophyceae** in 1998-1999 followed by *Spirogyra* (138.33 unit/l ±35.92), *Ulothrix* (137.67 unit/l ±77.36), *Oedogonium* (126.33 unit/l ±92.92), *Schizogonium* (116 unit/l ±68.15), *Chaetophora* (116 unit/l ±96.76), *Closterium* (108.67 unit/l ±109.00), *Vaucheria* (81.67 unit/l ±23.03), *Chlorella* (79.33 unit/l ±22.01), *Cosmarium* (52.33 unit/l ±70.12) and *Zynema* (35.67 unit/l ±31.37). In 1999-2000, it revealed *Closterium* was maximum (155.33 unit/l ±139.26) followed by *Spirogyra* (144 unit/l ±50.03), *Oedogonium* (135.33 unit/l ±51.32), *Cladophora* (134.67 unit/l ±60.21), *Schizogonium* (134 unit/l ±92.07), *Chaetophora* (129 unit/l ±136.54), *Ulothrix* (115.33 unit/l ±17.01), *Cosmarium* (111.33 unit/l ±98.29), *Chlorella* (91.33 unit/l ±111.09), *Vaucheria* (88.33 unit/l ±8.62) and *Zygnema* (84.67 unit/l ±53.95).

A great variation during different season was observed in **Chlorophyceae** at Pul-Jatwara. In summer season 1999-1999, *Spirogyra* was recorded highest (156 unit/l ±5.27) and *Zygnema* was found lowest (48 unit/l ±3.33). In monsoon season *Spirogyra* was noted maximum (97 unit/l ±5.16) and *Closterium, Cosmarium* and *Zygnema* were found as Nil. In winter season *Cladophora* was noted optimum 236 unit/l ±10.09 and *Zygnema* was found to be least (59 unit/l ±10.08). In summer season 1999-2000, *Spirogyra* was noted maximum (142 unit/l ±10.05) and *Schizogonium* was minimum (74 unit/l ±12.73). In monsoon season *Ulothrix* was highest (96 unit/l ±9.56) and *Chaetophora* and *Chlorella* were found as Nil. In winter season *Closterium* was noted optimum 315 unit/l ±12.11 and *Vaucheria* was observed least (96 unit/l ±8.33). The total highest reading for **Chlorophyceae** was observed in winter 1881 unit/l ±62.49, and lowest in monsoon season (599 unit/l ±40.56) for first year (1998-1999). In second year (1999-2000) maximum **Chlorophyceae** was detected during winter 2166 unit/l ±69.03 and minimum in monsoon 654 unit/l ±39.62 (Table 4B.33, Fig. 4B.41).

Table 4B.33. Quantitative Analysis for Seasonal Variation Among *chlorophyceae* of River Ganga at Sampling Station D (1998-2000)

	Seasons	Closterium	Cosmerium	Chaetophora	Chlorella	Cladophora	Oedogonium
1998 to 1999	Summer	108 ± 12.11	25 ± 9.22	102 ± 8.42	56 ± 5.31	128 ± 4.95	102 ± 14.91
	Monsoon	0 ± 0.00	0 ± 0.00	27 ± 2.64	80 ± 10.91	96 ± 9.88	48 ± 3.82
	Winter	218 ± 10.46	132 ± 5.91	219 ± 10.22	102 ± 12.49	236 ± 10.09	229 ± 10.25
	Average l^{-1}	108.67 ± 109.00	52.33 ± 70.12	116.00 ± 96.76	79.33 ± 23.01	153.33 ± 73.36	126.33 ± 92.92
1999 to 2000	Summer	92 ± 8.34	99 ± 6.49	115 ± 10.15	59 ± 6.28	116 ± 4.95	122 ± 5.29
	Monsoon	59 ± 12.55	23 ± 5.49	0 ± 0.00	0 ± 0.00	86 ± 4.77	92 ± 6.73
	Winter	315 ± 12.11	218 ± 11.06	272 ± 10.04	215 ± 8.22	202 ± 12.37	192 ± 12.09
	Average l^{-1}	155.33 ± 139.26	113.33 ± 98.29	129.00 ± 136.54	91.33 ± 111.09	134.67 ± 60.21	135.33 ± 51.32

Table 4B.33. (Contd.)

	Seasons	Schizogonium	Spirogyra	Ulothrix	Vaucheria	Zygnema	Total I $^{-1}$
1998 to 1999	Summer	68 ± 10.25	156 ± 5.27	105 ± 9.46	58 ± 6.82	48 ± 3.33	956 ± 38.97
	Monsoon	86 ± 3.52	97 ± 5.16	82 ± 8.22	83 ± 7.49	0 ± 0.00	599 ± 40.56
	Winter	194 ± 11.91	162 ± 8.48	226 ± 11.59	104 ± 11.06	59 ± 10.08	1881 ± 62.49
	Average I $^{-1}$	116.00 ± 68.15	138.33 ± 35.92	137.67 ± 77.36	81.67 ± 23.03	35.67 ± 31.37	1145.33 ± 37.42
1999 to 2000	Summer	74 ± 12.73	142 ± 10.05	128 ± 10.38	79 ± 13.22	130 ± 10.05	1156 ± 26.44
	Monsoon	88 ± 5.61	95 ± 9.55	96 ± 9.56	90 ± 8.24	25 ± 2.15	654 ± 39.62
	Winter	240 ± 14.22	195 ± 12.61	122 ± 12.60	96 ± 8.33	99 ± 7.43	2166 ± 69.03
	Average I $^{-1}$	134.00 ± 92.07	144.00 ± 50.03	115.33 ± 17.01	88.33 ± 8.62	84.67 ± 53.95	1325.33 ± 23.85

± = Standard Deviation.

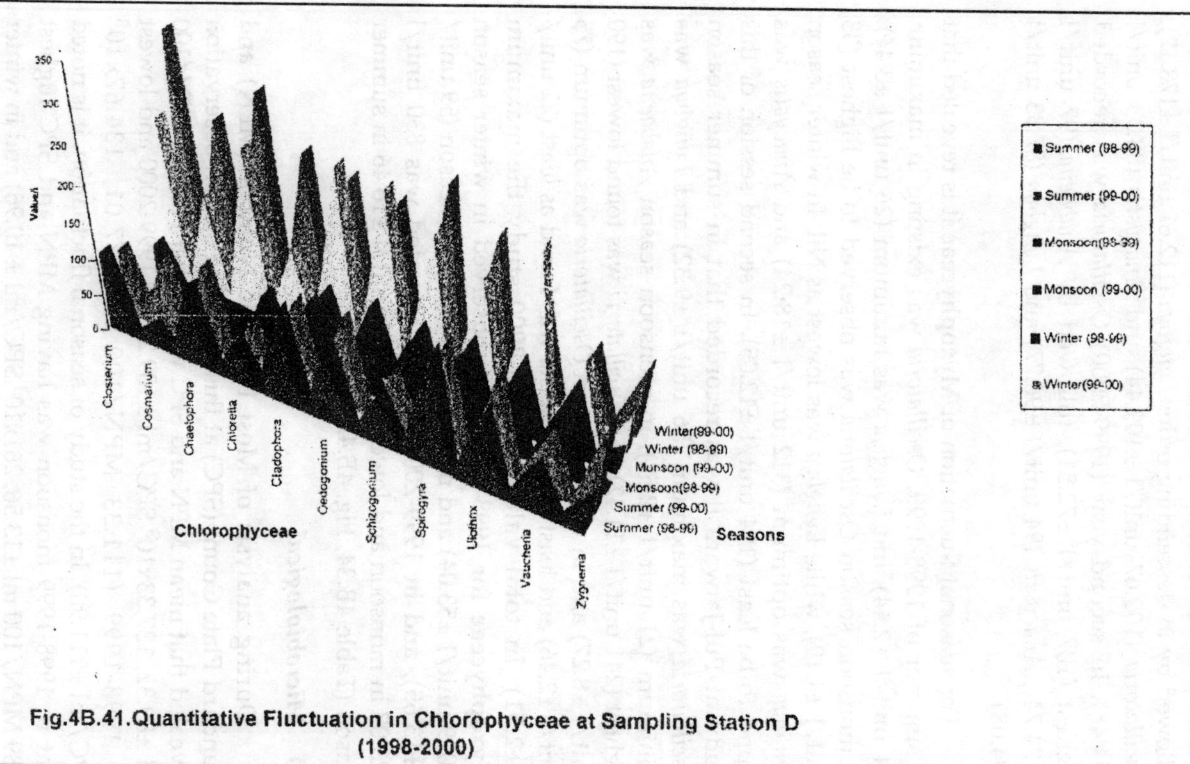

Fig.4B.41.Quantitative Fluctuation in Chlorophyceae at Sampling Station D (1998-2000)

In **Myxophyceae,** revealing average value, *Anabaena* was noted optimum (123 unit/l ±145.33) in first year (1998-1999), followed by in descending order *Lyngbya* (112.67 unit/l ±173.12), *Oscillatoria* (109.67 unit/l ±109.44) and *Anacystis* (88.33 unit/l ±94.48). In second year (1999-2000) *Oscillatoria* was recorded highest (107 unit/l ±72.51), followed by *Anabaena* (97 unit/l ±124.74), *Anacystis* (94 unit/l ±96.77) and *Lyngbya* (87.33 unit/l ±94.08).

For seasonal fluctuation of **Myxophyceae** it is revealed that in summer of 1998-1999, *Oscillatoria* was existing in maximum (61 unit/l ±2.44) and *Lyngbya* was minimum (26 unit/l ±3.44). In monsoon season *Oscillatoria* was observed to be highest (33 unit/l ±4.19), while *Lyngbya* was lowest as Nil. In winter season *Lyngbya* was optimum (312 unit/l ±18.24) and *Anacystis* was found to be least (194 unit/l ±12.05). In second session of this study at Pul-Jatwara it was recorded that in summer season *Oscillatoria* was maximum (56 unit/l ±9.32) and *Lyngbya* was minimum (21 unit/l ±2.51), in monsoon season *Anabaena* was highest (241 unit/l ±11.42) and *Oscillatoria* was found lowest (190 unit/l ±8.27) and in winter season *Oscillatoria* was optimum (75 unit/l ±2.46) and least was *Anabaena* was found as least (22 unit/ l ±2.51). In total value of 1998-2000 study the maximum **Myxophyceae** for 1998-1999 were observed in winter season (1029 unit/l ±53.04) and minimum in monsoon season (59 unit/ l ±13.65), and in 1999-2000 highest recorded was 830 unit/l ±23.07 in monsoon and lowest was 127 unit/l ±16.46 in summer season (Table 4B.34, Fig. 4B.42).

(b) Microbiological

During analysis of Most Probable Number (MPN) and Standard Plate Count (SPC) at this sampling station observation revealed that mean MPN and SPC was higher (119.33 MPN/100 ml ±89.32, 137.33x10^3 SPC/ml ±109.23) in 1999-2000 and lowest in 1998-1999 (111.33 MPN/100 ml ±77.01, 134.67x10^3 SPC/ml ±111.90). In the study of seasonal fluctuation, it is noted that in 1998-1999 monsoon was having MPN and SPC highest (189 MPN/100 ml ±13.56, 261x10^3 SPC/ml ± 10.96) and in winter season lowest (35 MPN/100 ml ±2.05, 48x10^3 SPC/ml ± 22).

Table 4B.34. Quantitative Analysis for Seasonal Variation Among *Myxophyceae* of River Ganga at Sampling Station D (1998-2000)

	Seasons	Anacystis	Anabaena	Lyngbya	Oscillatoria	Total l^{-1}
1998 to 1999	Summer	59 ± 6.29	67 ± 10.42	26 ± 3.44	61 ± 2.44	213 ± 18.48
	Monsoon	12 ± 1.11	14 ± 2.55	0 ± 0.00	33 ± 4.19	59 ± 13.65
	Winter	194 ± 12.05	288 ± 14.23	312 ± 18.24	235 ± 18.26	1029 ± 53.04
	Average l^{-1}	88.33 ± 94.48	123.00 ± 145.33	112.67 ± 173.12	109.67 ± 109.44	433.67 ± 14.56
1999 to 2000	Summer	22 ± 5.23	28 ± 1.06	21 ± 2.51	56 ± 9.32	127 ± 16.46
	Monsoon	204 ± 10.24	241 ± 11.42	195 ± 9.77	190 ± 8.27	830 ± 23.07
	Winter	56 ± 3.16	22 ± 2.51	46 ± 3.95	75 ± 2.46	199 ± 22.07
	Average l^{-1}	94.00 ± 96.77	97.00 ± 124.74	87.33 ± 94.08	107.00 ± 72.51	385.33 ± 8.18

± = Standard Deviation.

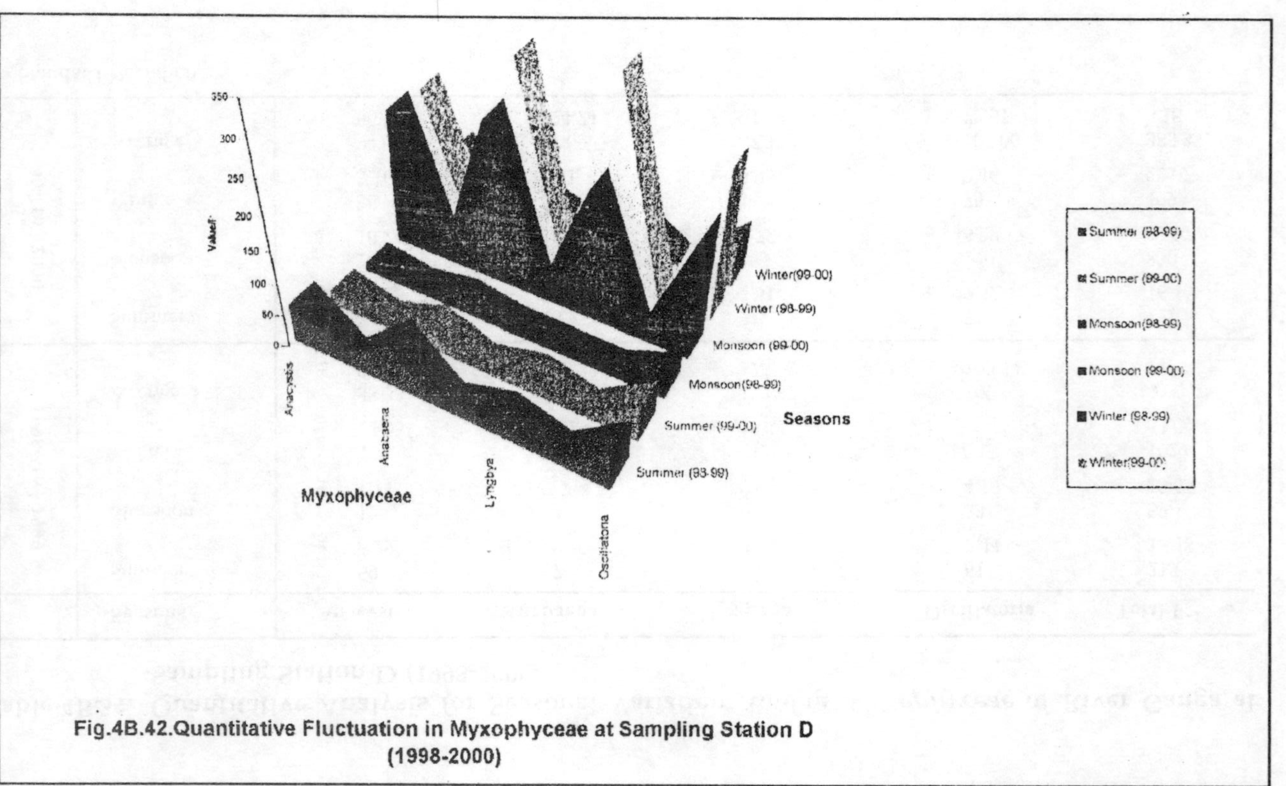

Fig.4B.42.Quantitative Fluctuation in Myxophyceae at Sampling Station D (1998-2000)

In 1999-2000 again monsoon season showed maximum MPN and SPC results (204 MPN/100 ml ±10.22, 256×10^3 SPC/ml ±12.27) and in winter season values were recorded as minimum (26 MPN/100 ml ±3.11, 41×10^3 SPC/ml ±4.24) (Table 4B.35 and Fig. 4B.43).

Table 4B.35. Quantitative Analysis for Seasonal Variation in Bateriological Parameters of River Ganga at Sampling Station D (1998-2000)

Bacteriology		MPN/100 ml	SPC/ml ($x10^3$)
1998 To 1999	Summer	110 ± 7.44	95 ± 7.17
	Monsoon	189 ± 13.56	261 ± 10.96
	Winter	35 ± 2.05	48 ± 3.22
	Average	111.33 ± 77.01	134.67 ± 111.90
1999 To 2000	Summer	128 ± 10.29	115 ± 7.97
	Monsoon	204 ± 10.22	256 ± 12.27
	Winter	26 ± 3.11	41 ± 4.24
	Average	119.33 ± 89.32	137.33 ± 109.23

± = Standard Deviation.

Bacterial cultures isolated from sampling station D shows that in 1998-1999 summer season, *Micrococcus luteus* was observed maximum (22×10^3/ml ±2.50) and minimum were *Proteus mirabilis, Aerobacter aerogens* and *Sporolactobacillus sp*. were found to be Nil. In monsoon season *Escherichia coli* was noted highest ($44x10^3$/ml ±1.59) and lowest was *Sporolactobacillus sp*. as Nil. In winter season optimum growth was observed for *Klebssilla pneumoniae* ($12x10^3$/ml ±2.50) and least growth were found to be Nil for *Sporolactobacillus sp*. In 1999-2000 summer season's observation revealed that growth of *Bacillus subtilis* was

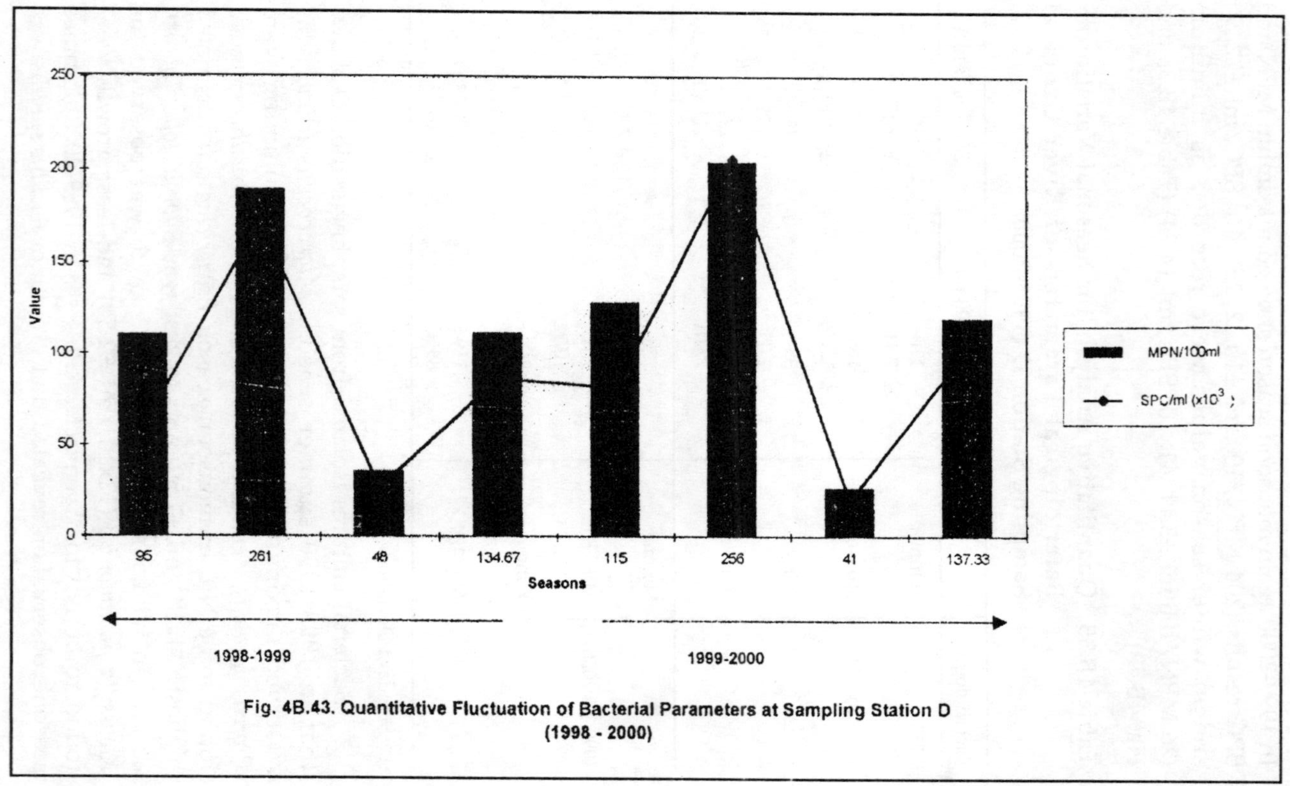

Fig. 4B.43. Quantitative Fluctuation of Bacterial Parameters at Sampling Station D (1998 - 2000)

optimum (26×10^3/ml ±1.59) and *Sporolactobacillus sp.* was observed least as Nil. In monsoon season growth of *Bacillus subtilis* was recorded as highest (40×10^3/ml ±2.22) and *Sporolactobacillus sp.* was found to be lowest as Nil. In winter season *Proteus mirabilis* showed optimum growth (12×10^3/ml ±2.00) and *Closterium, Salmonella typhii, Sporolactobacillus sp., Klebssilla pneumoniae, Staphylococcus epidermidis* and *Micrococcus luteus* were recorded as Nil.

On an average calculation for 1998-1999 *Escherichia coli* was observed highest (22×10^3/ml ±19.86) followed by *Micrococcus luteus* (18×10^3/ml ±2.58), *Bacillus subtilis* (15×10^3/ml ±8.89), *Klebssilla pneumoniae* (13×10^3/ml ±5.57), *Staphylococcus epidermidis* (12×10^3/ml ±12.58), *Pseudomonas aeruginosa* (11 $\times10^3$/ml ±15.37), *Staphylococcus aureus* (10×10^3/ml ±7.23), *Streptococcus facecalis* (9 $\times10^3$/ml ±8.33), *Clostridium perfringes* (9×10^3/ml ±8.89), *Proteus mirabilis* (7×10^3/ml ±8.33), *Salmonella typhii* (6×10^3/ml ±5.57), *Aerobacter aerogens* (4×10^3/ml ±4.04) and *Sporolactobacillus sp.* (absent). In 1999-2000 *Bacillus subtilis* was noted maximum (25×10^3/ml ±16.04) followed by *Escherichia coli* (24×10^3/ml ±14.57), *Micrococcus luteus* (15×10^3/ml ±10.82), *Pseudomonas aeruginosa* (12×10^3/ml ±9.07), *Staphylococcus aureus* (11×10^3/ml ±10.07), *Proteus mirabilis* (10 $\times10^3$/ml ±6.81), *Streptococcus facecalis* (10×10^3/ml ±7.09), *Staphylococcus epidermidis* (9×10^3/ml ±10.82), *Salmonella typhii* (7×10^3/ml ±8.08), *Aerobacter aerogens* (6×10^3/ml ±4.04), *Clostridium perfringes* (6×10^3/ml ±8.96), *Klebssilla pneumoniae* (4×10^3/ml ±4.00) and *Sporolactobacillus sp.* (absent) (Table 4B.36 and Fig. 4B.44).

Table 4B. 36. Isolation of Bacterial Culture from Sampling Station D (1998-2000)

Bacteria	1998 to 1999				1999 to 2000			
	Summer	Monsoon	Winter	Average	Summer	Monsoon	Winter	Average
Escherichia coli*	15 ± 1.50	44 ± 1.59	6 ± 0.50	22 ± 19.86	22 ± 1.50	39 ± 1.54	10 ± 1.00	24 ± 14.57
Pseudomonas aeruginosa*	4 ± 0.50	29 ± 2.51	1 ± 0.50	11 ± 15.37	8 ± 0.50	22 ± 1.50	5 ± 0.50	12 ± 9.07
Bacillus subtilis**	18 ± 0.50	22 ± 3.44	5 ± 1.20	15 ± 8.89	26 ± 1.59	40 ± 2.22	8 ± 0.50	25 ± 16.04
Proteus mirabilis**	0 ± 0.00	16 ± 0.56	4 ± 0.50	7 ± 8.33	2 ± 0.50	15 ± 1.50	12 ± 2.00	10 ± 6.81
Clostridium perfringes*	5 ± 0.50	19 ± 1.00	4 ± 1.50	9 ± 8.39	1 ± 0.50	16 ± 1.50	0 ± 0.00	6 ± 8.96
Streptococcus facecalis*	6 ± 1.50	18 ± 2.11	2 ± 0.51	9 ± 8.33	16 ± 1.52	11 ± 1.50	2 ± 0.50	10 ± 7.09
Aerobacter aerogens*	0 ± 0.00	8 ± 0.53	3 ± 0.50	4 ± 4.04	5 ± 0.50	10 ± 0.50	2 ± 0.50	6 ± 4.04
Salmonella typhii*	1 ± 0.52	12 ± 1.20	5 ± 1.50	6 ± 5.57	6 ± 1.00	16 ± 1.28	0 ± 0.00	7 ± 8.08
Staphylococcus aureus*	6 ± 0.54	18 ± 2.56	5 ± 0.50	10 ± 7.23	10 ± 1.50	22 ± 1.56	2 ± 0.50	11 ± 10.07
Sporolactobacillus sp.**	0 ± 0.00	0 ± 0.00	0 ± 0.00	0 ± 0.00	0 ± 0.00	0 ± 0.00	0 ± 0.00	0 ± 0.00
Klebssilla pneumoniae*	8 ± 0.50	19 ± 0.52	12 ± 2.50	13 ± 5.57	4 ± 0.50	8 ± 0.50	0 ± 0.00	4 ± 4.00
Staphylococcus epidermidis*	10 ± 0.53	25 ± 2.58	0 ± 0.00	12 ± 12.58	6 ± 1.50	21 ± 1.26	0 ± 0.00	9 ± 10.82
Micrococcus luteus**	22 ± 2.50	31 ± 3.41	1 ± 0.50	18 ± 12.58	9 ± 0.58	36 ± 1.52	0 ± 0.00	15 ± 10.82
Total ml $^{-1}$ ($\times10^3$)	95 ± 7.17	261 ± 10.96	48 ± 3.22	134.67 ± 111.90	115 ± 7.97	256 ± 12.27	41 ± 4.24	137.33 ± 109.23

± = Standard Deviation
Unit for value is ml $^{-1}$ ($\times10^3$)

* = Pathogen
** = Non-Pathogen.

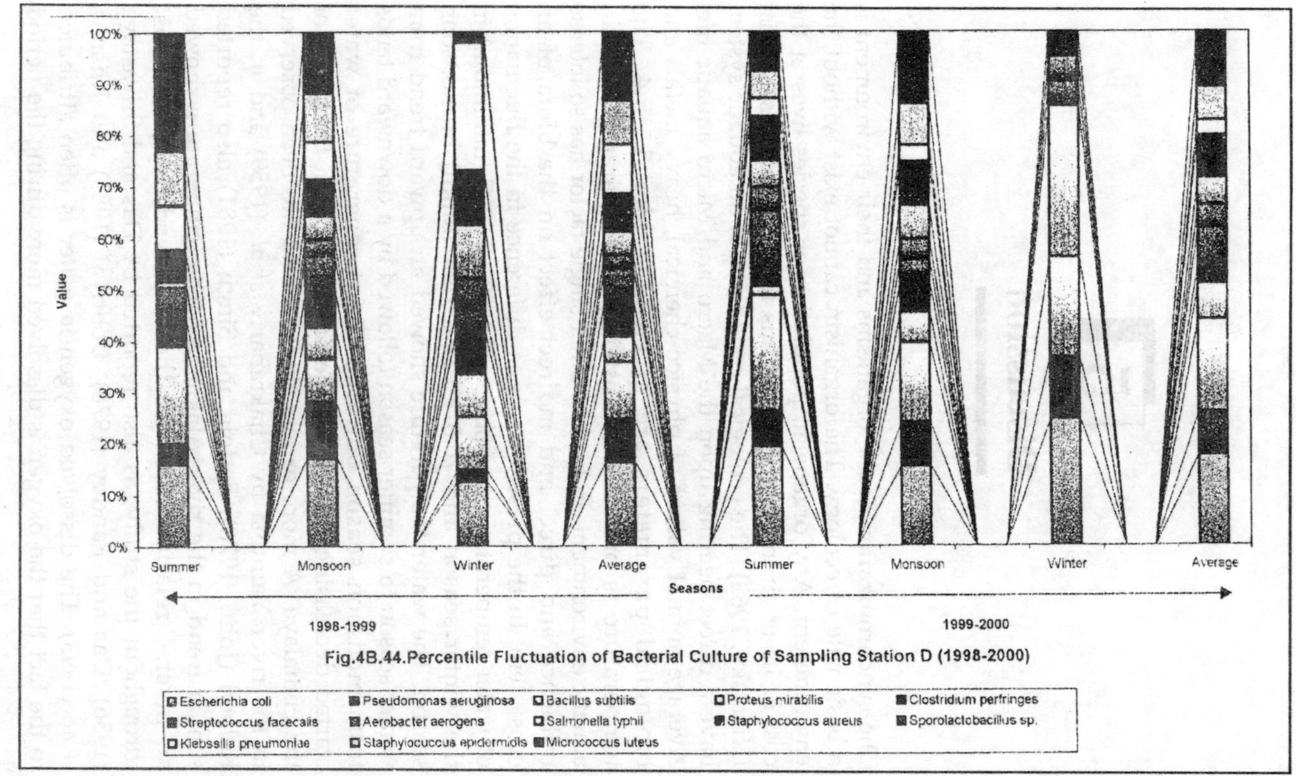

Fig.4B.44.Percentile Fluctuation of Bacterial Culture of Sampling Station D (1998-2000)

5

Discussion

The synchronization of the organisms and their environment is a basic rule of ecology. The organism cannot exist without the environment. Any organism, population or species lives at the cost of its environment. Without this relation it ceases to exist (Nikolsky, 1963). The ecological conditions of an aquatic system have a direct bearing upon the Macro and Micro aquatic life. Average annual values of physico-chemical, heavy metals and biological parameters are given in Tables 5.1-5.4. The temperature is one of the most important parameter in an aquatic environment. In fact, no other single factor has so intense influence and direct and indirect effect on the biota of an ecosystem. In the present study, a difference in the fluctuation of water temperature was observed 8.80 ±0.63°C (minimum) in the winter season and 20.37±0.22°C (maximum) in monsoon season. The water temperature showed an upward trend from winter season to summer season followed by a downward tends from monsoon season onwards. The temperature of water started decreasing due to the melting of snow at the peaks of the Himalaya. A more or less similar trend has been observed in the river Yamuna by Chakrabarty *et al.* (1959) and in the Kallayi (John, 1976). Badola and Singh (1981) also reported similar trend in the river Alaknanda. The temperature showed an inverse relationship with the dissolved oxygen almost throughout the study, as also reported by Das and Srivastava (1956), Das and Pandey (1978), Dobriyal (1985) and Khanna *et al.* (1993). The dissolved oxygen in water is often attributed to the fact that the oxygen is dissolved more during the period

Table 5.1. Average Annual Value of Physico-Chemical Parmeters

Physico-chemical Parameters	Sampling Station A		Sampling Station B		Sampling Station C		Sampling Station D		Average
	1998-1999	1999-2000	1998-1999	1999-2000	1998-1999	1999-2000	1998-1999	1999-2000	1998-2000
1. Water Temperature (°C)	15.80 ± 4.71	15.69 ± 5.97	15.67 ± 4.32	15.69 ± 5.97	15.41 ± 4.35	16.31 ± 6.48	15.61 ± 4.53	16.57 ± 6.27	15.84 ± 0.39
2. Turbidity (JTU)	197.61 ± 297.69	209.83 ± 316.50	199.93 ± 301.79	209.83 ± 316.50	198.05 ± 296.80	217.13 ± 329.08	198.35 ± 298.98	206.85 ± 309.65	204.70 ± 7.26
3. Conductivity (μmhos/cm²)	191.70 ± 127.77	204.52 ± 138.99	193.41 ± 130.35	204.52 ± 138.99	193.33 ± 130.59	208.41 ± 137.84	191.44 ± 127.32	206.70 ± 137.55	199.25 ± 7.39
4. Velocity (m/sec)	1.40 ± 0.60	1.58 ± 0.58	1.28 ± 0.58	1.58 ± 0.58	1.23 ± 0.57	1.50 ± 0.53	1.39 ± 0.58	1.55 ± 0.47	1.44 ± 0.14
5. Total Solids (mg/l)	1512.53 ± 2013.98	1563.84 ± 2025.55	1513.47 ± 2029.92	1563.84 ± 2025.55	1506.50 ± 2017.51	1566.81 ± 2055.99	1518.53 ± 2025.75	1615.81 ± 2130.54	545.17 ± 38.60
6. Total Dissolved Solids (mg/l)	373.01 ± 422.40	436.67 ± 501.13	378.42 ± 428.86	436.67 ± 501.13	376.84 ± 426.07	431.70 ± 486.69	373.67 ± 423.53	444.74 ± 508.07	406.47 ± 33.35
7. Total Suspended Solids (mg/l)	1139.00 ± 1595.12	1156.25 ± 1508.98	1135.16 ± 1604.33	1156.25 ± 1508.98	1129.65 ± 1594.86	1133.52 ± 1579.03	1144.86 ± 1605.24	1169.89 ± 1631.07	1145.57 ± 13.99
8. pH	7.32 ± 0.25	7.17 ± 0.13	7.33 ± 0.34	7.17 ± 0.13	7.28 ± 0.24	7.24 ± 0.01	7.32 ± 0.25	7.20 ± 0.18	7.25 ± 0.07
9. Free Carbon dioxide (mg/l)	2.98 ± 1.74	2.84 ± 1.69	2.98 ± 1.80	2.84 ± 1.69	2.97 ± 1.73	2.88 ± 1.65	2.99 ± 1.77	2.80 ± 1.74	2.91 ± 0.08
10. Dissolved Oxygen (mg/l)	9.49 ± 2.27	9.51 ± 1.94	9.64 ± 2.07	9.51 ± 1.94	9.51 ± 2.23	9.73 ± 1.89	9.55 ± 2.18	9.56 ± 1.99	9.56 ± 0.08

Table 5.1. (Contd.)

11. B.O.D. (mg/l)	2.25 ± 0.83	2.20 ± 0.59	2.25 ± 0.80	2.20 ± 0.59	2.26 ± 0.83	2.35 ± 0.41	2.26 ± 0.82	2.29 ± 0.53	2.26 ± 0.05
12. C.O.D. (mg/l)	8.67 ± 3.18	8.48 ± 2.26	8.69 ± 3.09	8.48 ± 2.26	8.69 ± 3.18	9.05 ± 1.57	8.69 ± 3.15	8.82 ± 2.04	8.69 ± 0.18
13. Total alkalinity (mg/l)	50.07 ± 15.75	50.74 ± 15.45	50.00 ± 15.94	50.74 ± 15.45	50.19 ± 15.59	51.70 ± 14.40	50.02 ± 15.76	51.93 ± 15.74	50.67 ± 0.77
14. Chloride (mg/l)	5.49 ± 5.27	5.32 ± 5.23	5.44 ± 5.18	5.32 ± 5.23	5.45 ± 5.20	5.34 ± 5.20	5.51 ± 5.29	5.31 ± 5.26	5.40 ± 0.08
15. Total Phosphate (mg/l)	0.07 ± 0.04	0.07 ± 0.05	0.07 ± 0.04	0.07 ± 0.05	0.07 ± 0.04	0.07 ± 0.05	0.07 ± 0.04	0.08 ± 0.06	0.07 ± 0.00
16. Total organic carbon (mg/l)	4.25 ± 0.94	4.28 ± 0.98	4.18 ± 1.04	4.28 ± 0.98	4.26 ± 0.95	4.50 ± 0.98	4.23 ± 0.91	4.43 ± 1.00	4.30 ± 0.11
17. Hardness (mg/l)	95.10 ± 10.91	100.32 ± 10.86	94.88 ± 11.13	100.32 ± 10.86	95.48 ± 11.35	98.86 ± 11.93	95.66 ± 11.71	97.57 ± 14.10	97.27 ± 2.31
18. Ions									
(a) Sodium (mg/l)	14.8285 ± 7.70	16.7344 ± 8.73	14.4650 ± 7.30	16.7344 ± 8.73	14.3100 ± 7.47	16.7557 ± 8.57	14.4894 ± 7.62	16.3936 ± 8.36	15.59 ± 1.15
(b) Potassium (mg/l)	1.8082 ± 0.70	1.8230 ± 0.72	1.8100 ± 0.74	1.8230 ± 0.72	1.8110 ± 0.72	1.8196 ± 0.78	1.7970 ± 0.72	1.8164 ± 0.76	1.81 ± 0.01
(c) Sulphate (mg/l)	20.6333 ± 3.06	20.4367 ± 3.20	21.0000 ± 3.18	20.4367 ± 3.20	20,4533 ± 2.76	19.8400 ± 3.22	20.2867 ± 2.57	19.8933 ± 2.84	20.37 ± 0.38
(d) Nitrate (mg/l)	0.0347 ± 0.02	0.0313 ± 0.03	0.0410 ± 0.03	0.0313 ± 0.03	0.0317 ± 0.02	0.0260 ± 0.02	0.0383 ± 0.03	0.0330 ± 0.03	0.03 ± 0.005

± = Standard Deviation.

Table 5.2. **Average Annual Value of Heavy Metals**

Seasons Parameters	Sampling Station A		Sampling Station B		Sampling Station C		Sampling Station D		Average
	1998-1999	1999-2000	1998-1999	1999-2000	1998-1999	1999-2000	1998-1999	1999-2000	1998-2000
Lead [Pb] (mg/l)	0.0723 ± 0.0081	0.0734 ± 0.0081	0.0722 ± 0.0082	0.0727 ± 0.0099	0.0684 ± 0.0064	0.0732 ± 0.0087	0.0704 ± 0.0072	0.0712 ± 0.0095	0.07173 ± 0.0017
Copper [Cu] (mg/l)	0.0174 ± 0.0011	0.0181 ± 0.0010	0.0173 ± 0.0013	0.0174 ± 0.0019	0.0530 ± 0.0312	0.0354 ± 0.0318	0.0176 ± 0.0007	0.0179 ± 0.0006	0.02428 ± 0.0132
Chromium [Cr] (mg/l)	0.0310 ± 0.0014	0.0319 ± 0.0042	0.0286 ± 0.0024	0.0322 ± 0.0058	0.0287 ± 0.0042	0.0297 ± 0.0043	0.0576 ± 0.0474	0.0672 ± 0.0629	0.03836 ± 0.0151
Cadmium [Cd] (mg/l)	0.0029 ± 0.0001	0.0032 ± 0.0006	0.0028 ± 0.0005	0.0031 ± 0.0008	0.0029 ± 0.0005	0.0033 ± 0.0006	0.0032 ± 0.0001	0.0032 ± 0.0005	0.00310 ± 0.0002
Zinc [Zn] (mg/l)	3.2088 ± 0.0317	3.2328 ± 0.0275	3.1021 ± 0.0864	3.1744 ± 0.0975	3.1679 ± 0.0491	3.2113 ± 0.0331	3.2201 ± 0.0377	3.2245 ± 0.0326	3.19274 ± 0.0433

± = Standard Deviation.

Table 5.3. Average Annual Value of Plankton

Plankton	Sampling Station A		Sampling Station B		Sampling Station C		Sampling Station D		Average
	1998-1999	1999-2000	1998-1999	1999-2000	1998-1999	1999-2000	1998-1999	1999-2000	1998-2000
1. Zooplankton l^{-1}	298.33	315.00	362.00	189.33	365.00	372.67	386.33	454.67	342.92
	± 234.11	± 252.47	± 316.50	± 181.57	± 436.83	± 281.54	± 504.94	± 595.78	± 77.89
2. Phytoplankton l^{-1}	2578.00	2118.33	2175.00	1925.33	2257.00	2531.33	1943.33	2206.00	2216.79
	± 1438.99	± 1608.06	± 1653.22	± 1708.94	± 2004.36	± 1467.94	± 1612.15	± 942.83	± 239.73

± = Standard Deviation.

Table 5.4. Average Value of Bacteriology

Bacteriology		MPN/100 ml	SPC/ml (x10^3)
Sampling Station A	1998-1999	77.33 ± 65.00	87.00 ± 77.13
	1999-2000	101.00 ± 90.50	102.67 ± 94.71
Sampling Station B	1998-1999	63.33 ± 52.52	55.67 ± 45.28
	1999-2000	72.67 ± 58.48	81.33 ± 72.67
Sampling Station C	1998-1999	94.00 ± 65.00	108.33 ± 88.60
	1999-2000	98.33 ± 77.22	111.67 ± 99.06
Sampling Station D	1998-1999	111.33 ± 89.32	134.67 ± 111.90
	1999-2000	119.32 ± 89.32	137.33 ± 109.23
Average		92.17 ± 19.48	102.33 ± 27.38

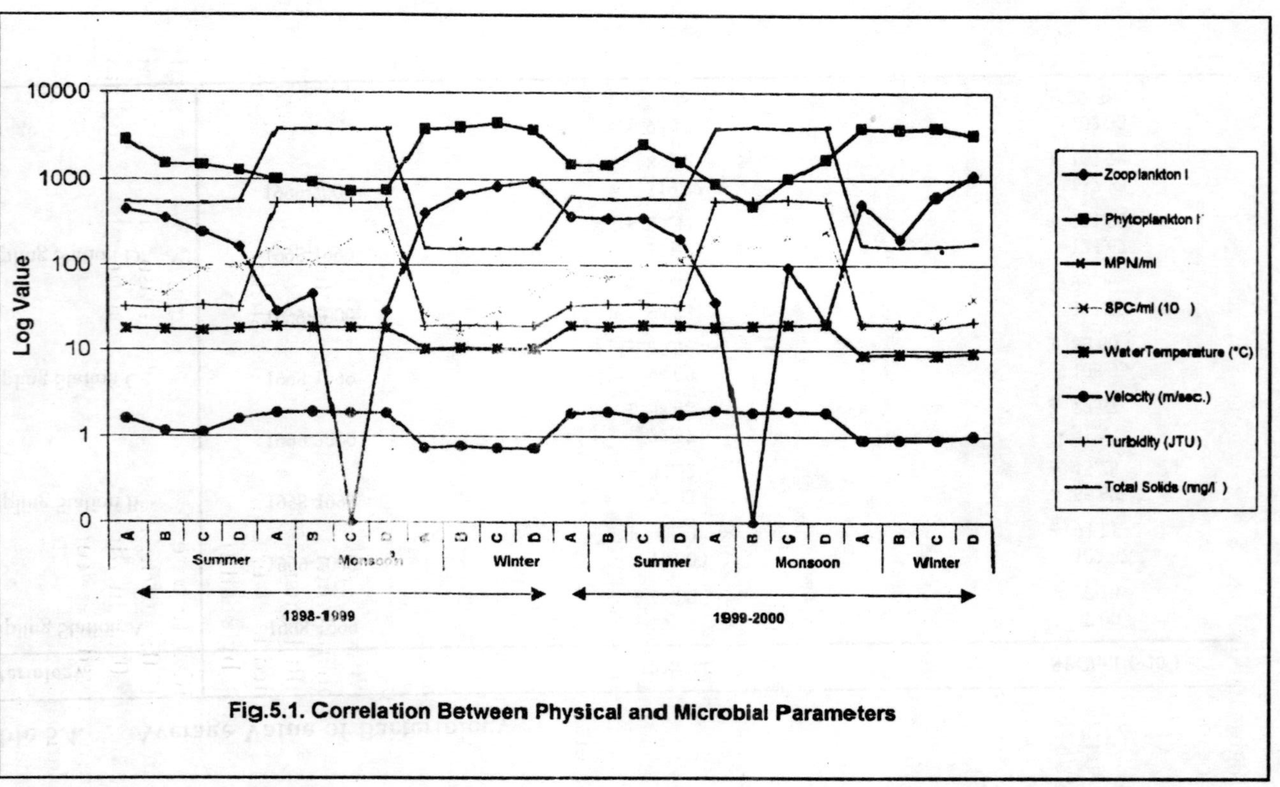

Fig.5.1. Correlation Between Physical and Microbial Parameters

of high catabolic activity by photosynthesis (if the temperature is not too high).

Maximum dissolved oxygen was recorded (11.74 mg/l ±0.43) in the winter season. The minimum value of dissolved oxygen (7.20 mg/l ±2.27) was observed in monsoon season. But the dissolved oxygen reduced gradually from summer onward due to turbidity, which retarded the photosynthetic activity of aquatic flora. This trend was also reported by Badola and Singh (1981) in the river Alkananda, Singh *et al.* (1982) in the river Nayar of Garhwal region and Khanna (1993) in the river Ganga. Concentration of dissolved oxygen is one of the most important parameter to indicate water purity and to determine the distribution and abundance of various algal groups. Many ecologists discussed the importance of dissolved oxygen for its role on various metabolic activities on different organisms. Roy (1955), Pahwa and Mehrotra (1966), Mishra and Yadav (1978), Verma (1978) and Bass and Harlet (1981) have discussed the seasonal fluctuations of dissolved oxygen and uniformly reported winter peaks in oxygen levels in their studies. This can be attributed to increased solubility of oxygen at lower temperature. It has been recommended that a minimum of 4 mg/l of dissolved oxygen should be maintained in water for healthy growth of fish and other microbial population. Different workers have pointed out various influencing factors on oxygen level that include discharges from industries, water current, velocity and biota. Adebisi (1981) observed high levels of dissolved oxygen on Ogun river in Nigeria during floods due to the aeration of river waters. Similar situation was reported in river Manjira when the river was in spate showing a value of 13 mg/l (Rajkumar, 1984). Highest values of dissolved oxygen were recorded in river Cauvery by Somashekar *et al.*, 1984, and in river Kosi by Bhatt *et al.*, 1984.

The BOD (biochemical oxygen demand) was observed maximum (3.03 mg/l±0.53) in monsoon season and minimum (1.38 mg/l ±0.10) in winter season. A negative relationship has been observed between BOD and DO contents (Fig. 5.2). A similar pattern has been reported by Khanna (1993). However, Verma *et al.* (1984) noted positive relationship of BOD and DO. BOD determination is a most useful technique to assess the level of organic pollution in river system. BOD levels were probably influenced by heavy metals with regard to seasonal fluctuations, peak values for BOD found to occur during monsoon and lower values during winter. On the contrary, Raina *et al.* (1984) reported peak values during winter in river Jhelum and Bhargava (1985) during summer for the river Yamuna. Martin (1970) designated water bodies moderately polluted if they exhibit BOD values above 8 mg/l.

COD (chemical oxygen demand) represents chemically oxidisable load of organic matter in water. It was noted highest (11.65 mg/l ±0.41) in winter season and minimum (5.31 mg/l ±2.15) in monsoon season. A negative relationship has been revealed between COD and water temperature (r= -0.76). River Ganga at Kanpur exhibits the highest value of COD, which was estimated to be 458 mg/l followed by river Sabarmati with a value of 417.0 mg/l.

Free carbon dioxide was observed maximum (5.05 mg/l ±0.20) in monsoon season due to higher turbidity and water temperature; but was recorded minimum (1.60 mg/l ±0.07) in winter season when the temperature of water decreased and turbidity was lowest. This phenomenon has been reported by Quadri and Shah (1984). Chakrabarty *et al.* (1959) also recorded the maximum free carbon dioxide in Jamuna during monsoon at Allahabad. Pahwa and Mehrota (1966) and Ray *et al.* (1966) have reported that the Ganga river contains maximum free carbon dioxide in rainy season at Allahabad. The free carbon dioxide is released during the decomposition of certain substances and metabolic activity of the living organisms. Since higher temperature accelerate the decomposition of organic substances as well as the respiratory activity of the biota, a direct relationship was established between the water temperature and free carbon dioxide (r= +0.61).

The dissolved oxygen and free carbon dioxide usually inversely related to one another because of the photosynthetic and respiratory activities of the biota (Hynes, 1970). The fluctuations in the dissolved oxygen content were mainly influenced by the factors like TOC, plankton and microbial activity.

Table 5.5. Trend Analysis of Bacterial Culture on Ganga River at Hardwar for Different Seasons

Bacteria	1998-1999												1999-2000												
	Summer				Monsoon				Winter				Summer				Monsoon				Winter				Average
	A	B	C	D	A	B	C	D	A	B	C	D	A	B	C	D	A	B	C	D	A	B	C	D	
Escherichia coli*	10	2	6	15	29	19	35	44	2	0	3	6	12	8	10	22	28	21	28	39	5	2	5	10	15
Pseudomonas aeruginosa*	2	0	2	4	20	22	29	29	0	0	0	1	5	2	5	8	36	12	33	22	1	0	0	5	10
Bacillus subtilis**	15	15	16	18	35	15	38	22	5	6	9	5	25	29	22	26	12	22	36	40	8	4	6	8	18
Proteus mirabilis**	1	0	2	0	5	2	6	16	0	0	0	4	0	2	4	2	6	8	5	15	0	0	0	12	4
Clostridium perfringes*	0	0	4	5	3	0	8	19	0	0	2	4	0	0	6	1	5	0	2	16	0	0	0	0	3
Streptococcus facecalis*	2	0	2	6	25	6	16	18	3	2	5	2	2	0	8	16	38	8	12	11	0	0	0	2	8
Aerobacter aerogens*	0	0	4	0	6	8	5	8	0	0	0	3	0	0	2	5	18	12	15	10	0	0	0	2	4
Salmonella typhii*	5	0	6	1	12	10	8	12	0	0	2	5	6	2	4	6	25	22	20	16	0	0	2	0	7
Staphylococcus aureus*	4	6	9	6	34	8	12	18	4	1	8	5	18	11	15	10	2	10	16	22	2	4	0	2	9
Sporolactobacillus sp.**	7	4	8	0	0	2	5	0	3	0	2	0	2	2	4	0	10	0	2	0	4	6	2	0	3
Klebssilla pneumoniae*	2	1	5	8	5	0	3	19	0	0	0	12	1	0	2	4	0	0	6	8	0	0	0	0	3
Staphylococcus epidermidis*	4	8	12	10	0	5	19	25	0	0	0	0	8	12	8	6	15	15	12	21	0	2	5	0	8
Micrococcus luteus**	8	10	13	22	0	8	21	31	10	7	0	1	3	2	5	9	11	29	31	36	0	2	2	0	11
Total	60	46	89	95	174	105	205	261	27	16	31	48	82	70	95	115	206	159	218	256	20	20	22	41	103

*Pathogens

**Non-pathogens.[

The water of the Ganga river at Hardwar becomes start turbid from summer season onward and in rainy season the river water was highly turbid due to flooded riverine condition. It showed that the turbidity and total solids were closely interrelated with one another and cause common effect upon the river and aquatic life as also stated by Verma *et al.* (1984). The lowest turbidity of water (19.11 JTU ±2.20) was recorded in winter season and highest (597.00 JTU ±26.4) noted in monsoon season. Similar pattern was also reported by Badola and Singh (1981) and Dobriyal *et al.* (1983) in the hill streams of the Garhwal Himalaya, and by Swarup and Singh (1979) in the lake Suhara. Dorris (1963) in the Mississippi noticed that the high water level is accompanied by less oxygen concentration brought about by the washing of organic matters and decreasing photosynthesis due to the turbidity. The turbidity is a striking characteristic to know the physical status of the rivers. The suspended particles, soil particles, discharged effluents, decomposed organic matter, total dissolved solids as well as the microscopic organisms increase the turbidity of water, which interferes with the penetration of light. The increased turbidity during monsoon season was attributed due to heavy soil erosion in the nearby catchments and massive contribution of suspended solids from sewage canals. Surface runoffs and domestic wastes mainly contribute to the increased turbidity of the river. Bhatt *et al.* (1984) observed low turbidity during November-February and an increase after May and reaches peak in August in river Kosi. He attributed that during monsoon months, the river water contained large amount of silt, fine and particles, organic matter and clay. In river Ganga and its tributaries, minimum values for turbidity were recorded in winter and summer seasons while, maximum in monsoon period (Bilgrami and Duttamunshi, 1985). Higher values were recorded at places where human activity was intense or at places where effluents from industries/farms are discharged into the river. High amount of silt carried due to fast currents of Kosi attributed mainly for the turbidity of this river (Bhatt *et al.*, 1984).

The optimum velocity (1.99 m/sec ±0.01) recorded in monsoon season and least (0.73 m/sec ±0.05) in winter season. The velocity started increasing after winter season due to melting of snow at the place of origin of the river. Again beyond

Table 5.6.A. Correlation Coefficient of Different Parameters

Columns: Physical (Water Temperature – Total Solids); Chemical (pH – Nitrate). Rows: Physical (Water Temperature – Velocity); Chemical (pH – Nitrate).

Prameters	Water Temperature (°C)	Turbidity (JTU)	Conductivity (µmhos/cm²)	Velocity (m/sec)	Total Solids (mg/l)	pH	Free Carbon Dioxide (mg/l)	Dissolved Oxygen (mg/l)	B.O.D. (mg/l)	C.O.D. (mg/l)	Total alkalinity (mg/l)	Chloride (mg/l)	Total Phosphate (mg/l)	Total organic carbon (mg/l)	Hardness (mg/l)	Sodium (mg/l)	Potasslum (mg/l)	Sulphate (mg/l)	Nitrate (mg/l)
Water Temperature (°C)		+0.53	+0.61	+0.91	+0.60	+0.66	+0.61	-0.71	+0.86	-0.76	+0.72	+0.59	+0.12	+0.15	+0.26	+0.38	+0.27	+0.46	+0.27
Turbidity (JTU)			+0.99	+0.69	+0.996	+0.577	+0.989	-0.925	+0.785	-0.883	-0.170	+0.995	+0.892	+0.904	+0.883	+0.974	+0.956	+0.972	+0.911
Conductivity (µmhos/cm²)				+0.753	+0.997	+0.608	+0.991	-0.953	+0.840	-0.917	-0.07	+0.995	+0.845	+0.860	+0.847	+0.955	+0.921	+0.948	+0.870
Velocity (m/sec)					+0.739	+0.656	+0.742	-0.807	+0.909	-0.864	+0.514	+0.730	+0.342	+0.380	+0.461	+0.584	+0.479	+0.610	+0.439
Total Solids (mg/l)						+0.617	+0.995	-0.945	+0.829	-0.914	-0.093	+0.998	+0.855	+0.868	+0.857	+0.955	+0.930	+0.962	+0.892
pH							+0.663	-0.662	+0.762	-0.763	+0.317	+0.630	+0.287	+0.342	+0.330	+0.429	+0.435	+0.557	+0.425
Free Carbon Dioxide (mg/l)								-0.949	+0.847	-0.932	-0.069	+0.997	+0.829	+0.851	+0.830	+0.935	+0.917	+0.957	+0.883
Dissolved Oxygen (mg/l)									-0.930	+0.961	-0.144	-0.945	-0.688	-0.704	-0.674	-0.846	-0.795	-0.851	-0.746
B.O.D. (mg/l)										-0.963	+0.430	+0.829	+0.439	+0.482	+0.480	+0.662	+0.587	+0.693	+0.541
C.O.D. (mg/l)											-0.227	-0.918	-0.602	-0.638	-0.624	-0.785	-0.736	-0.821	-0.700
Total alkalinity (mg/l)												-0.099	-0.580	-0.555	-0.451	-0.334	-0.448	-0.263	-0.441
Chloride (mg/l)													+0.852	+0.869	+0.850	+0.951	+0.931	+0.964	+0.893
Total Phosphate (mg/l)														+0.987	+0.943	+0.955	+0.979	+0.902	+0.936
Total organic carbon (mg/l)															+0.938	+0.960	+0.984	+0.908	+0.925
Hardness (mg/l)																+0.931	+0.932	+0.896	+0.900
Sodium (mg/l)																	+0.981	+0.946	+0.919
Potassium (mg/l)																		+0.957	+0.954
Sulphate (mg/l)																			+0.941
Nitrate (mg/l)																			

Table 5.6.B. Correlation Coefficient of Different Parameters

		Heavy Metals					Biological			
	Parameters	Lead (Pb) (mg/l)	Copper (Cu) (mg/l)	Chromium (cr) (mg/l)	Cadmium (Cd) (mg/l)	Zinc (Zn) (mg/l)	Phytoplankton I^{-1}	Zooplankton I^{-1}	MPN/100 ml	SPC/ml ($\times 10^3$)
Physical	Water Temperature (°C)	+0.910	+0.270	+0.260	+0.440	+0.020	-0.890	-0.710	+0.850	+0.710
	Turbidity (JTU)	+0.670	+0.028	-0.160	+0.419	+0.334	-0.691	-0.742	+0.799	+0.876
	Conductivity (μmhos/cm^2)	+0.728	+0.060	-0.119	+0.416	+0.314	-0.746	-0.785	+0.850	+0.902
	Velocity (m/sec)	+0.924	+0.120	+0.210	+0.494	+0.193	-0.885	-0.791	+0.882	+0.780
	Total Solids (mg/l)	+0.718	+0.055	-0.130	+0.423	+0.316	-0.738	-0.773	+0.834	+0.894
Chemical	pH	+0.651	+0.220	+0.088	+0.239	-0.011	-0.685	-0.479	+0.650	+0.588
	Free Carbon Dioxide (mg/l)	+0.722	+0.088	-0.142	+0.387	+0.282	-0.751	-0.774	+0.837	+0.891
	Dissolved Oxygen (mg/l)	-0.768	-0.117	+0.018	-0.278	-0.241	+0.849	+0.867	-0.877	-0.884
	B.O.D. (mg/l)	+0.864	+0.240	+0.072	+0.258	+0.088	-0.927	-0.862	+0.886	+0.825
	C.O.D. (mg/l)	-0.821	-0.197	+0.028	-0.277	-0.155	+0.879	+0.853	-0.879	-0.860
	Total alkalinity (mg/l)	+0.512	+0.303	+0.401	+0.028	-0.294	-0.526	-0.296	+0.348	+0.116
	Chloride (mg/l)	+0.710	+0.062	-0.142	+0.404	+0.303	-0.737	-0.769	+0.830	+0.892
	Total Phosphate (mg/l)	+0.331	-0.128	-0.308	+0.361	+0.442	-0.320	-0.474	+0.510	+0.676
	Total organic carbon (mg/l)	+0.349	-0.066	-0.307	+0.365	+0.441	-0.344	-0.496	+0.529	+0.689
	Hardness (mg/l)	+0.451	-0.103	-0.249	+0.546	+0.533	-0.394	-0.486	+0.602	+0.733
	Sodium (mg/l)	+0.558	-0.058	-0.218	+0.413	+0.415	-0.545	-0.661	+0.707	+0.819
	Potassium (mg/l)	+0.461	-0.066	-0.270	+0.359	+0.383	-0.467	-0.585	+0.621	+0.760
	Sulphate (mg/l)	+0.605	-0.043	-0.194	+0.434	+0.281	-0.630	-0.649	+0.701	+0.783
	Nitrate (mg/l)	+0.435	-0.131	-0.245	+0.365	-0.374	-0.446	-0.563	+0.585	+0.717
Heavy Metals	Lead [Pb] (mg/l)		+0.097	+0.007	+0.464	+0.062	-0.824	-0.669	+0.835	+0.739
	Copper [Cu] (mg/l)			-0.160	-0.091	-0.232	-0.208	-0.178	+0.199	+0.166
	Chromium [Cr] (mg/l)				+0.275	+0.166	-0.200	-0.152	+0.178	+0.058
	Cadmium [Cd] (mg/l)					+0.532	-0.323	-0.156	+0.529	+0.527
	Zinc [Zn] (mg/l)						+0.005	-0.169	+0.313	+0.405
Biological	Phytoplankton I^{-1}							+0.840	-0.855	-0.766
	Zooplankton I^{-1}								-0.778	-0.756
	MPN/100 ml									+0.958
	SPC/ml ($\times 10^{-3}$)									

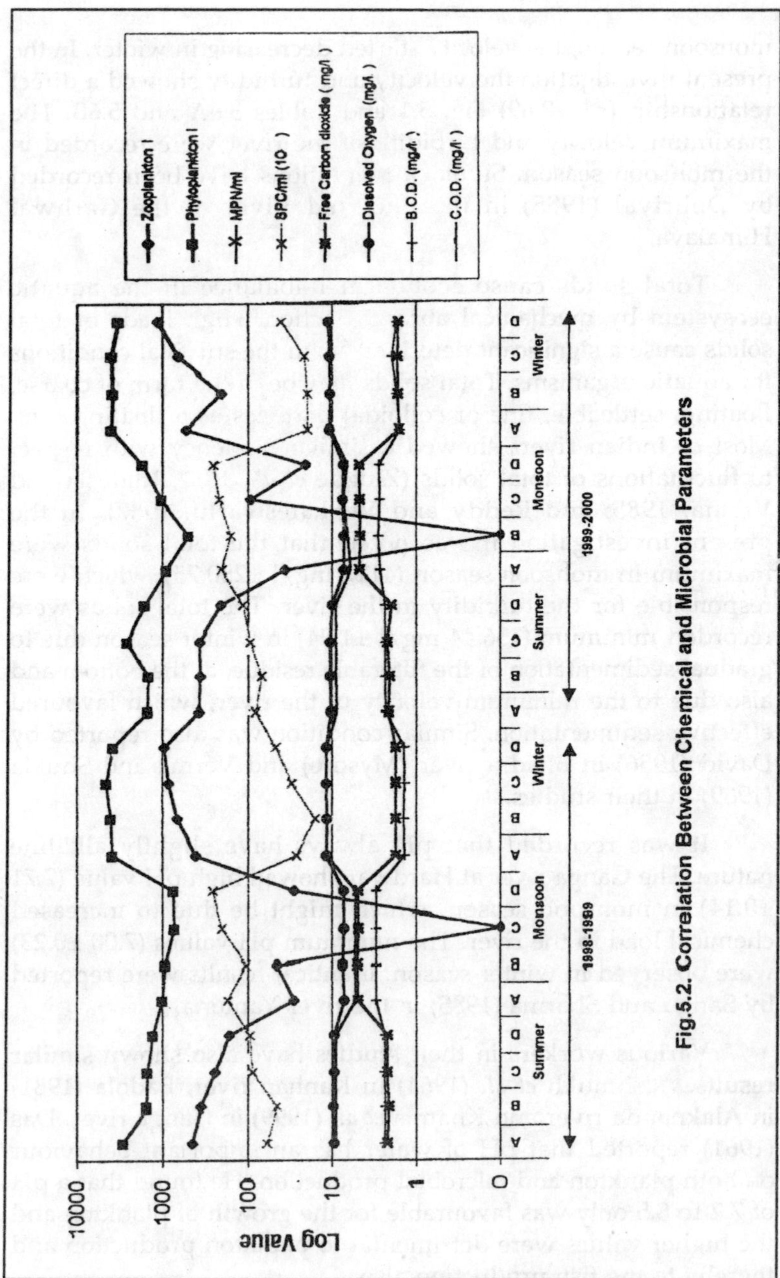

Fig.5.2. Correlation Between Chemical and Microbial Parameters

monsoon season, the velocity started decreasing in winter. In the present investigation the velocity and turbidity showed a direct relationship (r= +0.69) Fig. 5.1 and Tables 5.6A and 5.6B. The maximum velocity and turbidity of the river were recorded in the monsoon season. Similar observations have been recorded by Dobriyal (1985) in the Snowfed River of the Garhwal Himalaya.

Total solids cause ecological imbalance in the aquatic ecosystem by mechanical abrasive action. High loads of total solids cause a significant deterioration in the survival conditions for aquatic organisms. Total solids may be in the form of coarse, floating, settleable, fine or colloidal particles as a floating film. Most of Indian rivers showed a similar tendency with respect to fluctuations of total solids (Zingde *et al.*, 1930; Kudesia and Verma, 1985; and Reddy and Venkateswarlu, 1987). In the present investigation it was noted that the total solids were maximum in monsoon season (4117 mg/l ±250.75), which were responsible for the turbidity in the river. The total solids were recorded minimum (156.44 mg/l ±1.94) in winter season due to gradual sedimentation of the filterable residue, at the bottom and also due to the minimum velocity of the river, which favoured effective sedimentation. Similar condition was also reported by David (1956) in Bhadra river (Mysore) and Verma and Shukla (1969) in their studies.

It was recorded that pH always have slightly alkaline nature. The Ganga river at Hardwar showed high pH value (7.71 ±0.14) in monsoon season, which might be due to increased chemical load in the river. The minimum pH values (7.00 ±0.23) were observed in winter season. Identical results were reported by Sangu and Sharma (1985) in the river Yamuna.

Various workers in their studies have also shown similar results. Deshmukh *et al.* (1964) in Kanhan river, Badola (1981) in Alaknanda river and Khanna *et al.* (1999) in Ganga river. Das (1961) reported that pH of water has an important behaviour on both plankton and microbial production. He found that a pH of 7.2 to 8.5 only was favourable for the growth of plankton and the higher values were detrimental to plankton production and thereby to the fish production also.

Total alkalinity of water is a measure of weak acid present in it and of the cations balanced against them (Sverdrup *et al.*, 1942). The highest concentration (69.11 mg/l ±1.05) was observed in summer season and lowest (35.22 mg/l ±12.26) in winter season. Similar observations have also been made by Holden and Green (1960), Talling and Rzoska (1967) and Abdin (1948a). The alkalinity might be due to mixing of ashes, waste water from Ashrama and Bhimgoda nullah and washing activities. Venkateswarlu (1969) attributed that there is an indication to suggest that alkalinity and phosphate concentrations are affected directly by rainfall. Alkalinity was mainly due to bicarbonates throughout the year at all the four sampling stations. According to Hay and Anthony (1958) and Venkateswarlu and Jayanti (1968), the decomposition of the organic matter leads to the high alkalinity of the waters.

In study for Lead, Copper, Chromium, Cadmium and Zinc were taken for observation and revealed that all heavy metals are having partial or some how positive relation with velocity and negative relation with plankton life. Heavy metals get contaminated into aquatic systems as a result of various natural activities (weathering of soils and rocks from volcanic eruptions) and from a variety of human activities involving the mining, processing or use of metals or substances. Some metals such as manganese, iron, copper and zinc are essential micronutrients, while others such as mercury, cadmium and lead are not required even in small amount by any organisms. Virtually all metals, including the essential metal micronutrients, are toxic if exposure levels are sufficiently high. Cadmium, chromium, copper, lead, mercury, nickel and zinc are of particular relevance in fresh waters. The increased circulation of toxic metals in recent times resulted in the unavoidable build-up of such toxic substances in the human food chain.

Since heavy metals are rapidly absorbed to particulate materials (e.g. detritus, plankton, suspended sediments) and assimilated by living organisms. The concentration of metals dissolved in the water may give a highly misleading picture of the degree of metal pollution. Andren (1974) reported that about 67 per cent of the mercury in the Mississippi river water is associated with suspended sediments. Metals adhering particulate materials could be absorbed by an organism in the process of pumping water over its gills or in the process of filtering water to obtain food. Lloyd (1961) found that suspended Zinc was indeed toxic to Rainbow trout. There is paucity of information on the toxicity of suspended metals to most aquatic organisms.

The highest and lowest values of Chloride were found in monsoon season and winter season 11.58 mg/l ±0.26 and 1.72 mg/l ±0.23 respectively. It is one of the important chemical indicator of pollution. Chlorides are present in sewage, sewage effluents and farm drainage and remain unaltered during the purification of sewages. Rivers Yamuna (Sengar *et al.* 1985); Tungabhadra (Reddy and Venkateswarlu, 1987); Adyar (Govindan and Sundaresan, 1979); Jhelum (Raina *et al.* 1984) and Kshipra (Mishra and Saxena, 1984) also showed significant levels of chloride content.

The optimum concentration of Nitrates (0.07 mg/l ±0.01) was noted in monsoon season and lowest concentration (0.013 mg/l ±0.01) in the summer season. Nitrates concentration depends on the activity of nitrifying bacteria, which in turn get influenced by presence of dissolved oxygen. High concentrations of nitrates were observed after onset of rains (Prasad and Manjula, 1980; Venkateswarlu, 1969 Young *et al.*, 1973). On the contrary, Golterman (1975), Silabee and Larson (1982) observed higher nitrate levels after the rains. Raina *et al.*, (1984) while studying Jhelum river opinionated that the nitrate concentration increases during August and December owing to human activities and water flow. Zingde *et al.*, (1986) observed the lower nitrate values during May in Ambika and associated river estuaries.

The phytoplankton of many Indian rivers consists of diverse assemblage of major taxonomic groups. Many of these forms have different physiological and environmental requirements. The number, type and distribution of these organisms present in any aquatic habitat provide a clue on the environmental conditions prevailing in that particular habitat. However, it is seen that many environmental factors interact to provide conditions for the growth of plankton both specially and seasonally. Round (1970), while discussing the impact of various ecological parameters in the growth of different groups of plankton expressed problems of interpreting data particularly in case of biotic systems where changes in physico-chemical factors are marked with time. He emphasized the need for monitoring of a river system continuously at least for ten years to assess the importance of various chemical parameters in the growth and distribution of plankton and Khanna (1993) has also recommended same for the river Ganga at Hardwar.

Seasonal studies carried out by earlier authors on different rivers (Chakrabarty *et al.*, 1959) obtained peak values of total plankton during summer and minimum during monsoon. Roy (1955) and Venkateswarlu (1969), in contrast reported maximum values during winter and minimum during rainy season. Lakshminarayana (1965), Pahwa and Mehrotra (1966) observed seasonal maximum in river Ganga both in summer and winter. On the basis of the above observations by various authors, it can be supposed that the occurrence of seasonal maxima of total plankton depends more on the composition of the plankton and abundance of a particular group. One of the most important physical factors that influences plankton richness is the velocity of water current. It is stated that water current has inverse influence on plankton growth. Similar observations were drawn in present study and also reported by subsequent authors (Abdin, 1948; Allen, 1920; Blum, 1957; Butcher, 1924; Eddy, 1934; Kofoid, 1903; and Reinhard, 1931). Venkateswarlu (1969) also recorded an inverse correlation between water current and total algae. Current velocity has its own advantages and disadvantages as reviewed by Venkateswarlu (1969). Blum (1956) stated that water current caused mechanical danger to phytoplankton while Butcher, (1932) explained that there were difficulties to retain hold for the river bed algae with greater current speed.

In the present work it has been noted that the maximum total phytoplankton were 4528 unit/l ±1395.50 and zooplankton were 1134 unit/l ±105.27 in winter season. The minimum concentration of total phytoplankton 493 unit/l ±86.07 and zooplankton were nil in monsoon season. It was due to blanketing effect caused by the velocity. The study showed a dominance of phytoplankton, as also reported by Pahwa and

Mehrotra (1966) in the Ganga river. Badola and Singh (1981), and Singh *et al.* (1982) reported high values of plankton during January to March. However, the Ganga at Hardwar showed a single phytoplanktonic peak in January. Only one phytoplanktonic peak was also recorded by Chacko and Srinivasan (1955) in the Godawari, Chakrabarty *et al.* (1959) in the river Jamuna in July.

The number of plankton were also affected by the turbidity. The increased turbidity reduces the plankton production as also reported Das and Pathani (1978).

In the Ganga river at Hardwar the phytoplankton production was maximum in the winter when the temperature was low. It shows negative relationship between water temperature and plankton, which is also discovered, by Das and Srivastava (1956) and Khanna (1993). But Holmes and Whitton (1981) have reported that the density of plankton become greater in late summer in rivers of N.E. England.

According to Eddy (1934) the plankton production is mainly influenced by temperature. Kofoid (1908), Chakrabarty *et al.* (1959), and Pahwa and Mehrotra (1966) observed that in the warmer month the plankton production was medium. According to Bhowmick and Singh (1985), maximum density of phytoplankton was recorded in summer and lowest in Monsoon. Allen (1920) and Prasad (1956) have stated that temperature is the determining factor in the seasonal distribution of organisms. According to Holden and Green (1960), the temperature, when compared with certain other factors, has less significance in influencing the abundance of plankton. But Griffith (1955) could not establish any pronounced effect of temperature upon the total phytoplankton production.

The results indicated that the plankton were maximum in the winter months probably due to low temperature, high content of dissolved oxygen, low velocity.

It was observed that as soon as the water level, velocity and turbidity decreases, the water becomes clear. The diatoms and green algae appear on the surface of the stones, gravel and other substratum. Although during winter season the water temperature is lowest, even then the microbial life flourish very well (Fig. 5.1) because they receive maximum solar energy due to the high transparency of water.

However, a positive correlation between the pH and phytoplankton (r= +0.29) and zooplankton (r= +0.29) was established. In all the sampling stations the pH remain slightly alkaline.

The free carbon dioxide shows a positive relationship with the phytoplankton (r= +0.17) and with the zooplankton (r= +0.06) and also positive relationship between dissolved oxygen and phytoplankton (r= 0.33) and dissolved oxygen and zooplankton (r= +0.38).

The physico-chemical and biological parameters of all the four sampling stations were almost similar.

Due to the maximum velocity and volume of the river Ganga at Hardwar in monsoon season, the microbial populations were disturbed and decreased. Carpenter (1928) have reported that the plankton are sparse in swift running water. Allen (1920) has stated that water current is the chief factor influencing the plankton of the stream.

Study reveal that the value of Most Probable Number (MPN) coliform bacteria increases from summer season, maximum values was noted (204 MPN/100 ml ±10.22) in the monsoon season, and minimum was observed (9 MPN/100 ml ±2.66) in the winter season.

Since MPN and SPC have positive relationship, the higher value of Standard Plate Count (SPC) was recorded in the monsoon season (261×10^3 SPC/ml ±10.96) and minimum in the winter season (16×10^3 SPC/ml ±2.42).

Bissonnette *et al.* (1975), showed influence of environmental stress on enumeration or indicator bacteria from natural water; Bonda (1977), reported *Escherichia coli* as bacterial indicator of water pollution in 1981 he showed *Salmonella* and other pathogenic bacteria in fresh water. Dutka (1973), Dutka and Bell (1973) and Borrego *et al.* (1990) also reported that coliphages as an indicator of pollution. Borrego *et al.* (1987) reveled that *Escherichia coli, Salmonella sp.* and *Pseudomonas sp.* are indicator of fecal pollution in water source. The similar type of pollution indicators were found at all sampling stations and it revealed that the river Ganga at Hardwar is polluted due to coliform bacteria. It is also noted at sampling station D, that the *Escherichia coli* was higher than other station, as two nullahs are opening and mixing domestic and industrial effluents (towards upstream) in the maim stream.

On special notice it is also observed that *Staphylococcus aureus* which occurs on human being's skin, was comparatively high in summer season as number of pilgrim increases for Bathing activity on Bashakhi Festival and in monsoon season by Kanwarias (Lord Shiva Devotees).

Cherry *et al.* (1972) reported *Salmonella* as index of pollution of surface water. Kanner and Clark (1974) detected *Salmonella* and *Pseudomonas* in surface water. Goyal *et al.* (1977) noted occurrence and distribution of bacterial indicators and pathogens in canal communities along the Texas coast. Kuhen *et al.* (1997) reported biological fingerprinting of coliform bacterial populations. Comparison between polluted river and factory effluent. He noticed higher counts of bio-indicator on confluence of polluted water in river in down stream and Moriningo *et al.* (1990) established relationship between *Salmonella sp.* and indicator microorganisms in polluted natural waters.

The dominating bacteria of river Ganga at Hardwar are *Bacillus subtilis, Escherichia coli, Micrococcus luteus, Pseudomonas aeruginosa, Staphylococcus aureus, Staphylococcus epidermidis, Streptococcus facecalis, Aerobacter aerogens, Proteus mirabilis, Klepssilla pneumoniae, Clostridium perfringes* and *Sporolactobacillus sp.* (Table 5.5). The *Sporolactobacillus sp.* was least in the observation but, were appeared higher in summer season and monsoon season because many of the pilgrim worships Holy River Ganga with milk and curd.

Among 13 isolated and identified bacteria (plates 4.5-4.6.) some are noted pathogenic and others as non-pathogenic.

Escherichia coli, Pseudomonas aeruginosa, Clostridium perfringes, Streptococcus facecalis, Aerobacter aerogens, Salmonella typhii, Staphylococcus aureus, Klebssilla pneumoniae, and

Staphylococcus epidermidis were found in river Ganga at Hardwar as pathogens, where as the non-pathogenic bacteria were *Bacillus subtilis, Proteus mirabilis, Sporolactobacillus sp.*, and *Micrococcus luteus* (Table 5.5).

The value of Correlation Coefficient (r) between different parameters viz., physical-physical, physical-chemical, physical-Heavy metals, physical-microbiological, chemical-chemical, chemical-heavy metals, chemical-microbiological, heavy metals-heavy metals, heavy metals-microbiological and in between microbiological-microbiological are given in Tables 5.6A and Table 5.6B.

By the analysis for correlation coefficient of parameters, it extracted that there are highly significant correlation ($r \geq +0.9$) between Water temperature-Velocity (r= +0.91), Water temperature-Lead (r= +0.910), Turbidity-Conductivity (r= +0.99), Turbidity-Total Solids (r= +0.996), Turbidity-Free Carbon dioxide (r= +0.989), Turbidity-Chloride (r= +0.995), Turbidity-TOC (r= +0.904), Turbidity-Sodium (r= +0.974), Turbidity-Potassium (r= +0.956), Turbidity-Sulphate (r= +0.972), Turbidity-Nitrate (r= +0.911), Conductivity-Total solids (r= +0.997), Conductivity-Free Carbon Dioxide (r= +0.991), Conductivity-Chloride (r= +0.995), Conductivity-Sodium (+0.955), Conductivity-Potassium (r= +0.921), Conductivity-Sulphate (r= +0.948), Conductivity-SPC (r= +0.902), Velocity-BOD (r= +0.909), Velocity-Lead (r= 0.924), Total solids-Free Carbon Dioxide (r= +0.995), Total solids-Chlodide (r= +0.998), Total solids-Sodium (+0.955), Total solids-Potassium (r= +0.930), Total solids-Sulphate (r= +0.962), Total solids-Lead (r= +0.924), Total solids-SPC (r= +0.984), CO_2-BOD (r= +0.847), Free Carbon Dioxide-Chloride (r= +0.997), Free Carbon Dioxide-Sodium (r= +0.935), Free Carbon Dioxide-Potassium (r= +0.917), Free Carbon Dioxide-Sulphate (r= +0.957), Chloride-Sodium (r= +0.951), Chloride-Potassium (r= +0.931), Chloride-Sulphate (r= +0.964), Total Phosphate-TOC (r= +0.987), Total Phosphate-Hardness (r= +0.943), Total Phosphate-Sodium (r= +0.955), Total Phosphate-Potassium (r= +0.979), Total Phosphate-Sulphate (r= +0.902), Total Phosphate-Nitrate (r= +0.936), Total Organic Carbon-Hardness (r= +0.938), Total Organic Carbon-Sodium (r= +0.960), Total Organic Carbon-Potassium (r= +0.984), Total Organic Carbon-Sulphate (r= +0.908), Total Organic Carbon-

Nitrate (r= +0.925), Hardness-Sodium (r= +0.931), Hardness-Potassium (r= +0.932), Hardness-Nitrate (r= +0.900), Sodium-Potassium (r= +0.981), Sodium-Sulphate (r= +0.946), Sodium-Nitrate (r= +0.919), Potassium-Sulphate (r= +0.957), Potassium-Nitrate (r= +0.954), Sulphate-Nitrate (r= +0.941), MPN-SPC (r= +0.958).

In present study highly non-significant correlation (r≥ -0.75) were noted as follows—Water temperature-Turbidity-Dissolve Oxygen (r= -0.925), Conductivity-Dissolve Oxygen (r= -0.953), Total Solids-Dissolve Oxygen (r= -0.945), Total Solids-Chemical Oxygen Demand (r= -0.914), Free Carbon Dioxide-Chemical Oxygen Demand (r= -0.932), Dissolve Oxygen-Biological Oxygen Demand (r= -0.930), Dissolve Oxygen-Chloride (r= -0.945), Biological Oxygen Demand-Chemical Oxygen Demand (r= -0.963), Biological Oxygen Demand-Phytoplankton (r= -0.927), Chemical Oxygen Demand-Chloride (r= -0.918).

Remaining parameters was observed some how with positive and negative relationship (r≤ +0.9 or r≤ -0.9), which shows insignificant co-relation according to Tables 5.6A and 5.6B.

Few observations were noted about zero which shows no co-relation between those parameters e.g., pH-Zinc (r= -0.011), Dissolve Oxygen-Chromium (r= +0.018), Lead-Chromium (r= +0.007).

These isolated and identified pathogens (bio-indicators) in river Ganga at Hardwar tells very clearly that many effluents are contributing their pollution in main stream, those can be causing agents of many infection and diseases to the user of its water. This could be an alarming situation for pollution in river Ganga at Hardwar. Therefore, it is highly recommended that continues monitoring is most essential to maintain the purity of River Ganga at Hardwar as a Pilgrim and a pride for Hindus.

—

6

General Summary

The thesis exemplify the findings during research work conducted on "Microbial Ecology of River Ganga" during 1998-2000 and consists of 6 Chapters. 52 Tables, 95 Figures, including 2 Maps, and 10 Plates, support the text.

For this study, the water samples were collected from four spots [Sampling Station A (Har-ki-Pauri), Sampling Station B (Mayapuri Ghat), Sampling Station C (Muneshwar Ghat) and Sampling Station D (Pul-Jatwara)] over a distance of 8 km. Samples were analysed by the standard methods.

Sampling Station-A is situated in the north of Hardwar, a number of pilgrims take their holy bath and their number is increased manifolds at the time of bathing festivals. The ashes and bones are also dumped here. Before this station a cremation place is also situated and the effluents of Ashrams, Hotels and other houses also enter the Ganga at Pantdweep. Nullah of Bhimgoda and Kangra Mandir also open here. At the Sampling Station B, the overflow of Har-ki-Pauri comes to the main river. A dam is situated near this spot. The Sampling Station-C is situated downwards to Kankhal Bridge. The domestic sewage mixes upstream to the sampling point. The last Sampling Station-D, is situated at Pul Jatwara in Jwalapur, where two nullah are joining the main stream towards upstream of the water flow.

The river bottom at all the four sampling stations has stones and gravel covered by sand. The four spots were selected for this study due to differences in their physiography and other conditions.

The very first plain for river Ganga after running through Himalaya is at Hardwar. A summary of the findings is given below:

1. The water temperature of the Ganga at Hardwar ranged between 8.80°C±0.63 to 20.37°C ±0.22. The maximum water temperature started decreasing due to the melting of snow at the peaks of the Himalaya. The water temperature showed an upward trend from winter season to summer season followed by a downward trend from monsoon season onwards.

2. The turbidity in the river Ganga at Hardwar was lowest during winter season. From summer season onwards the water became turbid, due to melting of snow and rains. The maximum turbidity (597 JTU ±26.47) was observed in the monsoon season.

3. The conductivity of water is affected by the suspended impurities and also depends upon the amount of ions in the water. The highest conductivity (367 µmhos/cm^2 ±23.50) of the Ganga water was observed in monsoon season. From monsoon season onwards the conductivity decrease and minimum conductivity (99.33 µmhos/cm^2 ±2.18) was observed in winter season.

4. The velocity was found to be directly proportional to the flood level and also with gradient of the river stretch. The water level and its velocity started increasing form winter season onwards due to melting of snow at the place of origin of the river. The maximum velocity (1.99 m/sec ±0.01) of the Ganga river at Hardwar was recorded in monsoon season, and the minimum velocity (0.73 m/sec ±0.05) was observed in winter season.

5. Total solids may affect the water quality. Water with high total solids generally is of inferior potability. The total solids were maximum (4117 mg/l ±250.75) in monsoon season and it was minimum (156.44 mg/l ±1.94) in winter season.

Total dissolved solids were observed maximum (1090.33 mg/l ±98.25) in monsoon season and minimum (31.11 mg/l ±4.86) in winter season.

Total suspended solids were recorded maximum (3050.78 mg/l ±129.58) in the monsoon season and minimum (108.11 mg/l ±4.23) in winter season.

6. The pH of the Ganga river at Hardwar was slightly alkaline. It ranged from 7.71 ±0.14 to 7.00 ±0.23. However, any definite relationship between pH and plankton and microorganisms could not be established.
7. Free carbon dioxide in the Ganga water was invariably present throughout the year. It fluctuated from 1.60 mg/l ±0.07 to 5.05 mg/l ±0.20. The free carbon dioxide was found to be maximum in monsoon season and minimum in winter season.
8. The Ganga water contained highest dissolved oxygen during winter season, followed by a gradual decreased to its lowest values during monsoon season. The higher concentration of dissolved oxygen during winter season was probably due to low water temperature, no turbidity, and increased photosynthetic activities of the green algae found on the submerged stones and pebbles. The maximum (11.74 mg/l ±0.43) oxygen content of water was recorded in winter season and the minimum (7.20 mg/l ±2.27) in monsoon season. From monsoon season the water of the Ganga starts becoming turbid which reduced the photosynthetic activity of the algae and thus decreasing the oxygen concentration.
9. The BOD was maximum (3.03 mg/l ±0.53) in monsoon season and minimum (1.38 mg/l ±0.10) in winter season.
10. The COD ranged from 5.31 mg/l ±2.51 to 11.65 mg/l ±0.41. The minimum COD was recorded in monsoon season and maximum in winter season.
11. The total alkalinity was present throughout the year, which ranges from 69.11 mg/l ±1.05 to 35.22 mg/l

±12.26. The maximum total alkalinity was recorded in summer season and minimum in winter season.

12. The chloride was observed maximum (11.58 mg/l ±0.26) in summer season, and minimum (1.72 mg/l ±0.23) in winter season. The chloride showed significant relationship with ions.

13. The total phosphate was highest in monsoon season (0.14 mg/l ±0.01) and lowest in summer season (0.02 mg/l ±0.0).

14. Total organic carbon does not showed any significant correlation between plankton and microorganisms. The maximum TOC was observed (5.56 mg/l ±0.02) in monsoon season and minimum in summer season (3.16 mg/l ±0.05).

15. The hardness was higher in the monsoon season (113.83 mg/l ±2.96) and lower in summer season (83.33 mg/l ±1.50).

16. In all sampling station concentration of all ions (Sodium, Potassium Sulphate and Nitrate) were found in similar pattern and showed positive relation with bacterial and negative relation with plankton. This shows that these ions were required as micronutrient in the growth of bacteria. All the ions were higher in monsoon season and lower in summer season.

17. All the heavy metals (Lead, Copper, Chromium, Cadmium and Zinc) were maximum in monsoon season and minimum in winter season, except chromium and cadmium, which were minimum in the summer season. There was no significant relation between plankton and microorganism.

18. The velocity of water also has an impact upon the plankton and microorganisms of the Ganga. With the increase in velocity during monsoon season the plankton were washed away. There was a negative relationship between velocity and phytoplankton and velocity and zooplankton.

19. In all sampling stations, the minimum free carbon dioxide was found in winter when phytoplankton were maximum. A negative relationship observed between free carbon dioxide and phytoplankton and free carbon dioxide and zooplankton in winter season.
20. Plankton were found maximum in winter and minimum in monsoon season.
21. The genra of zooplankton and phytoplankton of the river Ganga at Hardwar at different study points are listed in the text.

    Total 8 genra of zooplankton were recorded. Rotifera group dominated among zooplankton. Total 32 genra of phytoplankton were observed. Among different groups of phytoplankton the Diatoms were found dominating.
22. The coefficient correlation among zooplankton and phytoplankton were found highly positive. The correlation of different parameters are given in the text.
23. Bacteria were found minimum in winters due to low temperature, more amount of dissolved oxygen, low velocity, minimum turbidity and maximum zooplankton of the water and bacteria were recorded maximum in monsoon.
24. Total 13 species of bacteria were recorded in the water of river Ganga at Hardwar. Among these species 9 were recorded pathogens (bio-indicators) and rest are non-pathogenic.
25. From the present study it may be concluded that:
    (i) There were only minor differences in physico-chemical and biological parameters of all the four sampling stations selected for this study.

    (ii) There was no pollution due to sewage or industrial effluents, but the major problem was ever increasing amount of silt.

    (iii) All the bacteria especially pathogens were not found in large number, but the presence of these

pathogenic bacteria is an alarm for increasing pollution status of river Ganga at Hardwar.

(iv) Beside this study, the regular monitoring of river Ganga at Hardwar is very important as a holy river and increasing instances of pollution. Thus, it is apparent that much attention should be paid on further studies of physico-chemical and microbiological parameters of river Ganga at Hardwar.

References

Abdin, G. (1948a). Physical and chemical investigations relating to algal growth in the river Nile. *Cairo. Bull. Inst. Egypt,* 29:20-24.

Abdin, G. (1948b). Seasonal distribution of phytoplankton and sessile algae in the river Nile. *Cairo. Ibid.,* 29:369-382.

Adebisi, A.A. (1981). The physico-chemical hydrology of a tropical seasonal river—Upper Ogun River (Nigeria). *Hydrobiol.,* 79(2):157-165.

Aggarwal, D.K., Gaur, S.D., Tiwari, I.C., Narayanaswami, M.S. and Marwali, S.M. (1976). Physico-chemical characters of the Ganga at Varanasi. *Indian J..Environ. Health,* 18:201-206.

Aggarwal, S.K. (1986). Ecology of Chambal River at Kota. *Acta Ecologia,* 8(1):13-19.

Alexander, W.B., Southgate, B.A. and Bassumdale, R. (1935). Survey of the river Tees. II the estuary chemical and biology Tech. Pap. *Wat. Poll. Res. No. 5.* HM SO London.

Alikunhi, K.H., Chaudhry, H. and Ramchandran, V. (1955). On the mortality of crop fry in the nursery ponds and role of plankton in the survival and growth, Indian J. Fish, 2:257-313.

Allen, W.E. (1920). A quantitative and statistical study of the plankton of the San Joaquin River and its tributaries in a near stockton California, 1913. *Univ. Calf. Publ. Zool.,* 22(1):1-24.

Anderson, G.C., Comita, G.W. and Engstrom, Heg, V. (1955). A note on the phytoplankton-zooplankton relationship in two lakes in Washington. *Ecology,* 36:757-759.

Andren, A.W. (1974). Ph.D. Dissertation, Florida State University Oceanography Department, Tallahassee, fla.

APHA (1980). *Standard Methods for the examination of water and waste water.* American Public Health Association, 1015 fifteen street NW Washington, 15:1-1134.

APHA, AWWA, WPCF (1985). *Standard Methods for the examination of water and waste water. 14th eds.* American Public Health Association, 1015 fifteen street NW Washington, 15:1-1134.

Atkin, W.R.G. and Harris, G.T. (1924). Seasonal changes in the water and heleoplankton of fresh water ponds Sampling Station C (Muneshwar Ghat). *Proc. Roly. Dub. Soc.* XVIII (N.S.): 1-21.

Badola, S.P. (1979). Ecological studies on the Ichythofauna of some fresh water resources of Garhwal Region. *D.Phil, Thesis, Garhwal University*.

Badola, S.P. and Singh, H.R. (1981). Hydrobiology of the river Alaknanda of Garhwal *Himalaya Indian J. Ecol.*, 8(2):269-276.

Badola, S.P. and Singh, H.R. (1981). Fish and fisheries of river Alaknanda. *Proc. Nat. Acad. Sci.*, 15 (B):133-142.

Badola, S.P. and Singh, H.R. (1982). Hydrobiology of river Alaknanda of the Garwal Himalayas. *Indian J. Eco.*, 8:269-276.

Baloni, M.C. and Sarkar, H.L. (1965). Some on the pollution of Yamuna river at Okhla water works intake Delhi. *Indian J. Environ, Health* 7:84-86.

Bass, D. and Harlet, R.C. (1981). Water quality of a South East Taxas stream. *Hydrobiol.*, 76:69-79.

Belsare, D.K. (1987). Limnology of protected water work in the Tropic. *Arch. Hydrobio. Beith Ergebn Limnol.*, 28:169-178.

Berner, L.M. (1951). Limnology of the lower Missouri River, *Ecology*, 32:1-12.

Bhargava, D.S. (1985). Water quality variations and control technology of Yamuna River. *Environ. Pollut.* (Series A), 37:355-372.

Bharati, A., Saxena, R.P. and Pandey, G.N. (1979). Physicological imbalances due to hexavalent Chromium on fresh water algae. *Indian J. Environ. Hlth.*, 2(3):234-243.

Bhatt, S.D., Bisht, Y. and Negi, U. (1984). Ecology of the Limnoflora in river Koshi of the Kumaun Himalaya (U.P.) *Proc. Indian. Nalu. S.C. Acad. B.*, 50(4):395-405.

Bhatt, S.D., Bisht, Y. and Negi, U. (1985). Ecology of the Limnoflora in the River Kosi of the Kumaun Himalaya (U.P.) *Proc. Natl., Sci.*, 50(4):395-405.

Bhatt, S.D., Bisht, Y. and Negi, U. (1985). Ecology and phytoplankton in river Kosi of the western Himalaya (U.P.) *Indian J. Ecol.* 12(1):141-146.

Bhowmick, B.N. and Singh, A.K. (1985). Phytoplankton population in relation to physico-chemical factors of River Ganga at Patna. *India J. Ecol.* 12(2):360-364.

Bilgrami, K.S. and Duttamunshi, J.S. (1985). Ecology of River Ganges (Patna-Farakka). *Technical Report, CSIR.*

Bissonnette G.K. Jezeski J.J., McFeters, G.A. and Stuart, D.G. (1975). Influence of environmental stress on enumeration of indicator bacteria from natural waters. *Appl. Microbiol.* **29,** 186-194.

Blum, J.L. (1957). An ecological study of the algae of the Saline River, Michigan. *Hydrobiol.,* 9(4):361-408.

Bonde, G.J. (1977). Bacterial indication of water pollution. In *Advances in Aquatic Microbiology* (Edited by Droop M.R. and Jannash H.W.). Vol. 1. Academic Press, London.

Bonde, G.J. (1981). *Salmonella* and other pathogenic bacteria. In *Water Supply and Health. Studies in Environmental Science 12* (Edited van Lelyveld, H. and Zoeteman, B.C.) Elsevier, Amsterdam.

Borchart, J.A. and Grahm Walton (1971). *Water quality in "Water quality and treatment" AWWA.* McGraw-Hill Book Company, 1-12.

Borrego, J.J., Cornax, R., Morningo, M.A., Mazanares, E.M. and Romero P. (1990). Colliphages as an indicator of fecal pollution in water. Their survival and productive infectivity in natural aquatic environments. *Water Res.,* 24(1):111-116.

Borrego, J.J., Morinigo, M.A., de Wicente, A., Cornax, R. and Romero, P. (1987). Colliphages as an indicator of fecal pollution in water. Its relationship with indicator and pathogenic microorganisms. Wat. Res. 21, 1473-1480.

Butcher, R.W. (1924). The plankton of the River Wharfe (Yorkshire). *Naturalist,* 175-180; 211-214.

Butcher, R.W. (1932). Studies on the ecology of river II. The microflora of rivers with special reference to the algae on the river-bed. *Ann. Bot. Lond.,* 46:813-861.

Butcher, R.W. (1946). Studies on the ecology of river IV. Algal growth in certain highly calcarious streams *J. Ecol.,* 33:268-283.

Butcher, R.W. (1949). Problems of distribution of sessile algae in running water. *Vernn., Inst. Ver. Limnol.,* 10:98-103.

Cairns, J.Jr. and Lanza, G.R. (1972). Pollution and Protozoan communities. In: *Water Pollution microbiology* edited by R. Mitchell, John Wiley & Sons, Inc. pp. 245-273.

Carpenter, K.E. (1928). Life in inland waters with special reference to animals. Sidwick & Jackson, London, 1-15;1-267.

Chacko, P.I. and Ganpati, S.V. (1949). Some observations on the Adyar River with especial reference to its hydrobiological conditions. *Indian Geogr. J.*, 24(3).

Chako, P.I. and Krishnamurthy, B. (1954). On the plankton of three freshwater fish ponds in Madras city, India. *Symp. Mar. and fresh water plank. Indo-Pacific Coun., Bangkok,* (UNESCO): 103-107.

Chako, P.I. and Srinivasan, R. (1955). Observations on the hydrobiology of the major rivers of Madras state, South India, *Fresh Biol. Stn. Madras,* 13:1-14.

Chandler, D.C. (1944). Limnological studies of western lake Eris. N. Relation of Limnological and climate factors to Phytoplankton of 1941. *Trans. Amer. Micros., Soc.,* 63:203-235.

Chakarbarty, R.D., Ray, P. and Singh, S.B. (1959). A quantitative survey of plankton and physiological conditions of the river Jamuna at Allahabad in 1954-1955. *Indian J. Fish,* 6(1):186-203.

Chandler, D.S. (1940). Limnological studies of western Eris. I: Plankton and certain physico-chemical Data of Bass Island region from September 1938 to November, 1939. *Ohio. J. Sci.,* 40:291-336.

Chattopadhya, S.N., Routh, T., Sharma, V.P., Arora, H.C., Gupta R.K. (1984). A short term study on the pollution status of Ganga river in Kanpur region. *Indian J. Environ. Hlth.,* 26(3):244-257.

Cherry, W.B., Hanks, J.B. Thomason, B.M., Murlin, A.M., Biddle, J.W. and Croom, J.W. (1972). Salmonellae as an index of pollution of surface water. *Appl. Microbiol.* **24,** 334-340.

Chopra, A.K., Patrick, Nirmal, J. (1994). Effect of domestic sewage on self-purification of Ganga water at Rishikesh. I. Physico-chemical parameters. *Ad. Bios.* Vol., 13(11):75-82.

Chopra, A.K. and Rehman, A. (1995). A study on self-purification of physico-chemical properties of Ganga canal water at Jwalapur, Hardwar. *Him. J. Env. Zool.* Vol. 9:11-13.

Choudhary, S.B. and Gouda, Rajashree (2000). Seasonal variability in the distribution of zooplankton in the costal water of Gopalpur: East coast of India. *Journal of Environment & Pollution* 7(3):175-183.

Clarke, F.W. (1924). The data Geochemistry. *Sthed-Bull.* U.S. Goel. Surv. p. 770.

Cosey, H. and Newton, P.V.R. (1973). The chemical composition and flow of the river Frome and its main tributaries. *Fresh wat. Biol.,* 3:317-333.

Crayton, W.M. and Sommerfield, M.R. (1979). Composition and abundance of phytoplankton in tributaries of the lower Colarado river, Canyon region. *Hydrobiologia*, 66(1):81-93.

Daniel, C.C., Wilder, H.B. and Weiner, M.S. (1982). *Water quality of the French Broad river North Corolina. An analysis of data collected at Marshal 1958-77*. Wat. Supp; Pap. 2185 A-D US Geol Surv. Govt. Print. Press Washington, D.C.

Das, H.B., Kalita, H., Saikia, L.B., Borah, K. and Goswami, C.C. (1989). Potability of waters of Arunachal Pradesh. JIWWA, 21:191.

Das, S.M. (1961). Hydrogen ion concentraiton, plankton and fish in fresh water Eutrophic lakes of India, *Nature* 191(4787):511-12.

Das, S.M. and Pathani, S.S. (1978). A study on the effect of lake ecology on productivity of Mahaseer (*Tor tor and Tor putitora*) in Kumaon lakes, India, *Matsya*, 4:25-31.

Das, S.M. and Upaddhyaya, J.C. (1979). Studies on qualitative and quantitative fluctuations of plankton in two Kumaon lakes, Nanital and Bhimtal (India). *Acta. Hydrobiol.*, 21(1):9-17.

Das, S.M. and Srivastava, V.K. (1956a). Some new observations on fresh water plankton Pt. I, plankton of fish tanks of Lucknow *Sci. and Cult.*, 21(8):466-467.

Das, S.M. and Srivastava, V.K. (1956b). Quantitative studies on fresh water plankton, Part II. Correlation between plankton and hydrological factors. *Proc. Nat. Acad. Sci. India*, 26B:243-254.

Das, S.M. and Srivastava, V.K. (1959). Studies on fresh water plankton III. Quantitative composition and seasonal fluctuations in plankton components, *Proc. Nat. Acad. Sci. India*, 29:174-189.

Deshmukh, S.B., Phadke, N.S. and Kothandaraman, V. (1964). Physico-chemical characteristics of Kanhan River Water (Nagpur). *Indian J. Environ. Hlth.*, 6(3):181-186.

Dobriyal, A.K. (1985). Ecology of Limnofauna in the small streams and their importance to the village life in Garhwal Himalaya. *Uttar Pradesh J. Zool.* 5(2):139-144.

Dobriyal, A.K. and Singh, H.R. (1981). Diurnal variation in some aspects of limnology of the river Mandakini from the Garhwal Himalaya, *Uttar Pradesh J. Zool.*, 1:16-18.

Dobriyal, A.K. and Singh, H.R. (1987). A case study on the origin of rhithroplankton in Garhwal Hill streams. *Agri. Biol*........; 3(2):104-106.

Dobriyal, A.K. and Singh, H.R. and Bist, K.L. (1983). Diurnal variation in Hydrobiological parameter of two Hill streams of Garhwal Himalaya, *India. Uttar Pradesh J. Zool.* 3:30-34.

Dorris, T.C. (1963). Limnology of the middle Mississippi River. IV Physical and chemical limnology of river and chute. *Limnol. Occanogr.*, 8:79-99.

Dutka, B.J. (1973). Coliforms are an inadequate index of water quality. *J. Envir. Hlth.*, 36, 39-46.

Dutka, B.J. and Bell J.B. (1973). Isolation of Asalmonellae from moderately waters. *J. Wat. Pollut. Control Fed.*, 45, 316-323.

Eddy, S. (1934). A study of the fresh water plankton communities. *Illinois, Biol, Monoar* 12(4):1-93.

Ellis, M.M., Westfall, B.A. and Ellis, M.D. (1946). Determination of water quality. *US. Fish and Wild. Ser. Res. Dept.* 28:122-136.

Fesher, N.S., Michel, B. and Lovis, T. Jean. (1984). Accumulation and toxicity of Cd, Zn, Ag and Hg in four marine phytoplankters. *Mar. Ecol. Prog. Ser.*, 18:201-213.

Freiser, H. and Fernado, Q. (1966). *Ionic equilibria in analytical chemistry*, Willey and Sons, New York.

Gainey, P.L. and Lord, T.H. (1957). Microbiology of water and sewage. *Preuticle Hall Inc. XII*, p. 430.

Galtsoff, P.S. (1924). Liminlogical observation in the Upper Mississippi, 1921. *Fish Bull* V.S., 39:347-438.

Ganapati, S.V. (1940). The ecology of the temple tank containing a permanent bloom of *Microcystis aerguinosa* (Kulx) *Hevfr. J. Bombay Nat. Hist. Soc.*, 42(1):65-77.

Ganapati, S.V. and Alikunhi, K.H. (1950). Factory Effluents from the mettar chemical and Industrial Corporation Ltd. Metteur Dam. Madras and their pollutional effects on the fishes of River Cauvery. *Proc. Nut. Inst. Sci.* India. 16(3):189.

Garner, J.H., Brown, F.W. and Lovett, M. (1936). Chemical and biological survey of the river Holme. *Rep. W. riding Riv. Bd.*

George, M.G., Qasim, S.A. and Siddiqui, A.Q. (1966). A limnological survey of the River Kali with reference to fish mortality, *Envir. Health*, 8(4):262-269.

Ghousuddin, M. (1934). Notes on the algal flora of the river Mossi, Hydarabad (Deccan) *J. Qsm. Univ.*, 1-3.

Ghosh, B.B. and Basu, A.K. (1968). Observation on estuarine pollution of the Hooghly by the effluents from a chemical factory complex at Rishra, W.B. (India). *Env. Hlth.*, 10:204-218.

Golterman, H.L. (1975). Chemistry. In: *River ecology* edited by B.A. Whitton. Black-well Scientific Publication, Oxford, London, Edinburgh, Melbourne pp. 39-80.

Gopal, B., Goel, P.K., Sharma, K.P. and Trivedy, R.K. (1981). Limnological study of fresh water reservoir, Jamwa Ramgarh (Jaipur), *Hydrobiologia*, 83:283-294.

Govindan, V.S. and Sundaresan, B.B. (1979). Seasonal succession of algal flora in polluted region of Adyar River. *Indian J. Environ. Hlth.*, 21(2):131-142.

Goyal, S.M., Gerba, C.P. and Melnick, J.L. (1977). Occurrence and distribution of bacterial indicators and phatogens in canal communities along the Texas coast. *Appl. Envir. Microbiol.* 34, 139-149.

Greenfield, R.E. (1925). Comparison of chemical and bacteriological examination made on the Illinois river during a season of low and season of high water. *1923-24 Illionis State Wat. Surv. Bull* 10:9-33.

Griffith, R.S. (1955). Analysis of plankton yield in relation certain physical and chemical factors of lake Michigan. *Ecology*, 36(4):543-552.

Gupta, S.L. (1989). Interactive effects of nitrogen and copper on growth of cyanobacterium *Microcystis*. *Bull Environ. Contam Toxicol.*, 42(2):215-222.

Hawkes, H.A. (1975). *Studies in Ecology*, Volume-2, River Ecology, Oxford, 312-374.

Hays, E.R. and Anthony, E.H. (1958). *Limnol. Oceanogr.*, 3(3):297-307.

Ho, Y.B. (1984). Zn and Cu concentrations in *Ascophyllum nodosum* and *Fucus vasialosus* (Phaeophyta, Fucales) after transplantation to an estuary contaminated with mine waters. *Conserv. Recycl.*, 7:267-372.

Holden, J.M. and Green, J. (1960). Hydrobiology and plankton of the river Sokoto. *J. Anim. Ecol.*, 29(1):65-84.

Holmes, N.T.H., and Whitton, B.A. (1981). Phytoplankton of four rivers, the Tyne, Wear, Tees and Swale. *Hydrobiologia*, 80:111-127.

Horned, D.A. (1982). *Water quality of Neuse river north Carolina-variability, pollution loads and long term trends.* Water supply paper 2185 A-1. U.S. Geol. Surv. Govt. Printing Office, Washington, D.C.

Hynes, H.B.N. (1970). *The ecology of running waters.* Liverpool University Press, Liverpool, 4th impression: 1-555.

Iyengar, M.O.P. and Venkataraman, G. (1951). The ecology and seasonal succession of the River Cooum at Madras with special reference to the diatomaceae. *J. Madras Univ.*, 21:140-192.

Jackson, D.F. (1961). Comparative studies on phytoplankton photosynthesis in relation to total alkalinity. *Verh. Int. ner. Limnol.,* 14:125-133.

Jameel, A.A. (1998). Physico-chemical studies in Uyyakondan channel water of river Cauvery. *Poll. Res.* 17(2):111-114.

Jeejabai, N. and Rajendran, S. (1982). Phytoplankton constituents as indicator of water quality—A study of Adyar River. *Proc. Muragappa Chettiar Symposium* 267-272.

John, V. (1976). Hydrobiological studies on the river Kallayi in Kerala. *Indian J. Fish.,* 23:72-85.

Joshi, B.D., Pathak, J.K., Singh, Y.N., Bisht, R.C.S. and Joshi, P.C. (1993). On the physico-chemical characteristics of river Bhagirathi in Uplands of the Garhwal Himalaya. *Him. J. Env. Zool.,* 7:64-75.

Joshi, B.D., Bisht, R.C.S. and Semval, V.P. (1995). Primary productivity in western Ganga Canal at Hardwar. *Indian. J. Ecol.,* 22(2):123-126.

Joshi, B.D., Bisht, R.C.S. and Joshi, Namita (1996). Planktonic population in relation to certain physico-chemical factors of Ganga canal at Jwalapur (Hardwar). *Him. J. Env. Zool.* Vol. 10:75-77.

Joshi, C.B. (1996). Hydro-Biological profile of River Sutlej in its middle stretch in Western Himalayas, *U.P.I.Zooi.,* 16(2):9-103.

Kanner, B.A. and Clark, H.P. (1974). Detection and enumeration of *Salmonella* and *Pseudomonas aeruginosa J. Wat. Pollut. Control Fed.* 46, 2163-2171.

Kant, S. and Anand, V.K. (1978). Interrelationship of phytoplankton and Physical factors in Mansar lake, Jammu, (J & K). *Indian J. Ecol.,* 5(2):134-140.

Kant, S. and Kachroo, P. (1973). Limnological studies in Kashmir lakes. *Proc. Indian Natn. Sci. Acad.,* 39(B), 5:632-645.

Kaushik, N.K. and Prashad, D. (1964). Coliform periodicity in water of Yamuna at Wazirabad, Delhi. *Indian J. Environ. Hlth,* 2:118-124.

Khanna, D.R. (1993). *Ecology and Pollution of Ganga River.* Ashish Publishing House, Delhi:1-241.

Khanna, D.R., Badola, S.P., Singh, H.R. and Dobriyal, A.K. (1992). Observations on Seasonal Trends in Diatomic Diversity in the river Ganga at Sapt-sarovar, Hardwar. *Recent research in coldwater: fisheries,* ed. By K.L. Sehgal, Today and Tomarrow Printers and Publishers, New Delhi. pp. 99-107.

Khanna, D.R., Malik, D.S. and Badola, S.P. (1997). Population of green algae in relation to physico-chemical factors of the river Ganga at Lal-Ji-Wala, Hardwar. *Uttar Pradesh J. Zool.* 17(3):237-240.

Khanna, D.R., Malik, D.S. and Rana, D.S., (1997). Phytoplanktonic communities in relation to certain physico-chemical parameters of Ganga canal at Hardwar. *Him. J. Env. Zool.*, 12:193-197.

Khanna, D.R., Badola, S.P. and Dobriyal, A.K. (1993). Plankton ecology of the river Ganga at Chandighat, Haridwar. *Advance in limnology*. Ed. By H.R. Singh, Narendra Publishing House, New Delhi. 171-174.

Khanna, D.R., Malik, D.S., Seth, T.R. and Rupendra, (1999). Correlation between abiotic factors and planktonic population in river Ganga at Rishikesh (U.P.). *Sus. Eco. Sys. & Env.*, Ed. By D.R. Khanna, A. Gautam and A. Goutam, Published by ASEA, Rishikesh. pp. 69-76.

Kofoid, C.A. (1908). The Plankton of the Illinois river 1894-1899, with introductory notes upon the Illinois river and its basin Part III. Constituent organisms and their seasonal distribution. *Bull. Ili. State. Lab. Nat. Hist.*, 8:1-354.

Kuehn, I., Allestam, G., Engdahl, M., Stenstroem, T.A., Morris, R., Grabow, W.O.K. and Jofre. J., (1997). Biochemical fingerprinting of coliform bacterial populations—comparisons between polluted river water and factory effluents. *Health Related Water Microbiology*. pp. 343-350.

Kudesia, V.P. and Verma, S.P. (1985). A study of industrial pollution on Kali. *River J. Env. Sci.* 1(2):41-49.

Kumar, A. (1995). Studies on pollution in river Mayurakshi in South Bihar *Indian. J. Env. Poll*, 2(1): 21-26.

Lakshminarayanan, J.S.S. (1965). Studies on phytoplankton, of the River Ganges, Varanasi, India. *Hydrobiological*, 25:119-164.

Lickens, G.E. (1961). Primary production of aquatic ecosystems. In: *The primary productivity of the Biosphere*, edited by H. Lieth and R.H. Wittakar, Springer verlgg, New York.

Lloyd, R. (1961). The toxicity of mixture of zinc and copper sulphates to rain bow trout (*Salmo gaidnerii)* Richardson. *Ann. Appl. Biol.*, 49:535-538.

Martin, J.H. (1970). Phytoplankton—Zooplnakton relationships in Narranganset Bay IV. The seasonal importance of grazing. *Limnol. Ocenogr;* 15:413-418.

Martin, R.S., Gates, W.H., Tobin, R.S., Grantham, D., Sumarah, R., Wolfe, P. and Forestall, P. (1982). Factors affecting coliform bacteria growth in distribution system. *Am. Water work Ass. J.*, 74:34-37.

Mathur, R.P. (1982). *Water and waste water testing*. Nem Chand and Bros. Publishers, Roorkee., 1-54.

Mishra, G.P. and Yadav, A.K. (1978). A comparative study of physico-chemical characteristics of rivers and lakes in Central India. *Hydrobiologia,* 59(3):275-278.

Misra, N. and Saxena, A.B. (1984). The effect of sewage with special reference to aquatic insects in the River Kshipra (India). *Int. J. Environ. Studies,* 23:191-208.

Motwani, M.P., Banerji, S.M. and Karam Chandani, S.J. (1956). Some observation on the pollution of river Sone by the factory effluent of Rohtas Industries at Dalmia nagar (Bihar). *Indian, J. Fish.,* 3:334-367.

Mazhar, M.D. and Kapoor, C.P. (1992). Limnological Studies on Doraniu river at Bareilly (U.P.). *J. Fresh Water Bio.* 4(2):155-158.

Mohanty, R.C. (1981). Water quality studies of some water bodies of Bhubaneswar, *Ph.D. Thesis,* Utkal University, pp. 1-240.

Moriningo, M.A., Cornax, R. Munoz, M.A. and Romero, P. (1990). Relation between *Salmonella* spp. and indicator microorganisms in polluted natural waters. *Water Res.,* 24(1):117-120.

Nautiyal, P. (1990). *Ecology of the Ganga river system in the upland of Himalaya Environment*. Resource and Development:69-73.

Nautiyal, P., Dobriyal, A.K., Chaukiyal, D.C. and Singh, H.R. (1986). Quality and potability of the Alakananda water at Srinagar—Gharwal (U.P., India). A note on Bacterial density as an indicator of pollution. *Poll. Res.,* 5(3-4):143-145.

Needham, J.G. and Needham, P.R. (1972). *A guide to the study of freshwater Biology*. Holden-Day I.N.C. San Francisco. Calif., 94(3):1-108.

Nikolsky, G.V. (1963). *The Ecology of fishes*. Academic Press, London, 1-352.

Oliff, W.O. (1959). Hydrobiological studies on the Tugela river system. Part I. The main Tugela River, *Hydrobiologia,* 14:281-370.

Pahwa, D.V. and Mehrotra, S.N. (1966). Observations on fluctuations in abundance of plankton in relation to certain hydrological conditions of River Ganga. *Proc. Nat. Acad. Sci.* 36B(2):157-189.

Palmer, C.M. (1962). *The effect of Pollution on river algae.* U.S. Dept. Hlth. Edn., and Welfare Public Health Ser., Rober. A. Taft Sanitary Engineering Center, Cincinnati, Ohio.

Pandey, B.N., Kumar, K., Lala, A. and Das, P.K.L. (1993). A preliminary Study on the physico-chemical quality of water of the river Koshi. *J. Ecobiol.,* 5:237-339.

Pathak, J.K. (1991). Some aspects of the ecology of river Sarju and Gomti of Kumaun Himalaya. *Ph. D. Thesis*. Kumaun University Nanital.

Patrick, R. (1978). Effects of trace metals in the aquatic ecosystem. *Am. Scientist*, 66(2).

Pokhriyal, R.C., Dobriyal, A.K. and Singh, H.R. (1983). Seasonal variation in potamological parameter of the steam Khandagad. *Uttar Pradesh J. Zool.*, 3:70-72.

Prasad, B.N. and Yashpal, S. (1982). On diatoms as indicators of water pollution. *J. Indian Bot. Soc.*, 61:326-336.

Prasad, B.N. and Manjuala, B.N. (1980). Ecological study of Blue-green algae in River Gomti. *Indian J. Environ. Hlth.*, 22(2):151-168.

Prasad, R.R. (1956). Further studies on the plankton of the inshore waters of Mandapam. *Indian J. Fish.*, 2(1):1-41.

Probost, J.L. (1986). Dissolved and suspended matter, transported by the Girau River (France): Mechanical and chemical erosion rate in a Calerious Mollasse basin. *J. Des. Sciences Hydrobiologioue.*, 31:61-79.

Pruthi, H.S. (1933). Studies on the bionomics of fresh water of India-I. Seasonal changes in the physical and chemical conditions of water of the tank in the Indian Museum compound. *Int. Rev. Hydrobiol.*, 28:46-67.

Quadri, M.Y. and Shah, G.M. (1984). Hydrobiological features of Hoassar. A typical wetland of Kashmir-I, Biotope. *Indian, J. Ecol.*, 2(2):203-206.

Raghunathan, M.B. (1978). Studies on seasonal Tanks in Tamil Nadu I. Chambarambakkam Tank. *Indian J. Zool.*, 19(2):81-85.

Rai, L.C. (1978). Ecological studies of algal communities of the Ganga river at Varanasi *Indian J. Ecol.*, 5:1-6.

Rai, L.C. (1984). Ecological studies of algal communities of Ganga river at Varanasi *Indian Jr. Environ. Hlth.*, 26(3):187-201.

Raina, V., Saha, A.R. and Ahmed, S.R. (1984). Pollution studies on river Jhelum-I: An assessment of water quality. *Indian J. Env. Hlth.*, 26:187-210.

Rajkumar, B. (1984). *Ecological studies in the river Manjira (AP) with special reference to its lithophytic flora.* Ph.D. Thesis, Osmania University, Hyderbad.

Ramesh, N.B., Srikanth, R. and Mulimohan Rao, A. (1992). Evaluation of water quality of three rivers of Andhra Pradesh. *Poll. Res.* 11(4):209-212.

Rao, C.B. (1955). On the distribution of algae in a group of six small ponds II. Algal periodicity. *Indian J. Ecol.*, 43:291-308.

Rao, D.S. and Govind, B.V. (1964). Hydrobiology of Tungbhadra reservoir *Indian J. Fish.*, 2(1):321-344.

Ray, H.K. (1955). Plankton ecology of the river Hoogly at Palta. *Ecology*, 36(2):169-175.

Ray, P., Singh, S.B. and Sehgal, K.L. (1966). A study of some aspects of the river Ganga and Jamuna at Allahabad (U.P.) in 1958-59. *Proc. Nat. Acad. Sci. India*, 36 B(3):235-272.

Reinhard, E.G. (1931). The plankton ecology of the Upper Mississippi, Minneapolis to Winona. *Ecol. Monogr.*, 1:395-464.

Reddy, P.M. and Venkateswarlu, V. (1987). Assessment of water quality and pollution in the River Tungabhadra near Kurnool, (A.P.) *J. Environ. Biol.*, 8(2):109-119.

Richardson, R.E. (1921). Changes in bottom and shore fauna of the middle Illinois river and its connecting lakes since 1913-1915, as a result of the increased southward of sewage pollution. *Illinois Nat. Hist. Surv. Bull.*, 14:33-75.

Robinson, J.M. and Kaller, E.C. (1976). A comparison of the water characteristics of four northern west Virginia rivers. *Jr. Proc. w.va. Acad. Sci.*, 48(1):1-67.

Roy, H.K. (1955). Plankton ecology of the river Hoogly at *Palta*. *Ecology*, 36(2):169-175.

Ross, Fedrick. C. (1983). *Introductory Microbiology*. Charles E. Merill Publishing Company. A Bell & Howell Company, Columbus, Ohoio, pp. 1-615.

Round, F.E. (1970). *The biology of algae*. London, Edward Amold.

Sangu, R.P.S. and Sharma, K.D. (1985). Studies on Water Pollution on Yamuna River at Agra. *Indian J. Environ. Hlth.*, 27(3):257-261.

Saxena, K.K. and Chauhan, R.R.S. (1993). Physico-chemical aspects of pollution in river Yamuna at Agra. *Poll. Res.* 12(2):101-104.

Saxena, K.L., Chakaraborty, R.N., Khan, A.Q. and Chattopadhay, S.N. (1966). Pollution studies of river Ganga near Kanapur, *Indian J. Env. Health.*, 8:270-385.

Sengar, R.M.S., Sharma, K.D. and Pathak, P.D. (1985). Studies on distribution of algal flora in polluted and non-polluted regions in Yamuna River at Agra (U.P.). *J. Indian Bot. Soc.* 64:365-376.

Sharma, Hardeep Rai, Chhetry, Deepak, Kaushik, Anubha and Trivedi R.C. (2000). Veriability in organic pollution of river Yamuna in Delhi. *Journal of Environment & Pollution*, 7(3):185-188.

Sharma, R.C. (1985). Seasonal abundance of phytoplankton in the Bhagirathi river, Garhwal Himalaya. *Indian J. Ecol.*, 12(1):157-160.

Sharma, R.C. (1986). Effect of physico-chemical factors on Benthic fauna of Bhagirathi river Garhwal Himalaya, *Indian J. Ecol.* 13(1):133-137.

Sharma, S.D. and Pandey, K.S. (1998). Pollution status on Ramganga river at Moradabad. *Poll. Res.* 17(2):201-209.

Silabee, D.G. and Larson, G.L. (1982). Water quality of streams in the Great Smoky Mountains—National Park USA *Bydrobiol.*, 89(2):97-115.

Singh, K.P. (1960). The Chlorophyceae of Kumaon Hills, U.P. India I *Agra Univ. J. Res. (sceince).* 9(2):211-216.

Singh, H.P., Choudhry, M. and Kalekaur, V. (1982). Seasonal and Diurnal changes in physico-chemical features of the River Brahamputra at Gauhati, *Ind. J. Zool.*, 2(82):77-84.

Singh, H.R. (1991). *Advances in limnology.* Narendra Publication House, New Delhi, p. 322.

Singh M. (1967). Phytoplankton and water temperature, Silicon and pH in a lake in Delhi. *Phykos,* 7:120-128.

Sivakumar, A.A., Thirumathal, K. and Aruchami, M. (1987). Effect of pollution on the physico-chemical qualities of the River Amaravati (Tamil Nadu). *Proc. Nat. Conf. Environ. Impact Bio.*, Madras.

Shivsubramani, R. and Mahadevan, A. (1995). Water quality of river (River Suruliyar) in Tamil Nadu. *Poll Res.*, 14(1):72-82.

Shrivastava, V.K. Shrivastava, G.K. and Shrivastava, J.K. (1996). Phytoplankton productivity and physico-chemical properties of Repti river. *Ecol. Env. and Conc.*, 2(183-185).

Somashekar, R.K. (1984). Phytoplankton constituents as indicator of water quality. A study of the River Cauvery. *Int. J. Environ. Studies,* 24:115-123.

Sverdrup, H.H., Johnson, M.W. and Fleming, R.H. (1942). *The Oceans, their physics, chemistry and general biology.* Prentice Hall, Inc., New York.

Swarup Krishan and Singh, S.R. (1979). Limnological studies of Suraha lake (Ballia). *J. Inland Fish Soc. India,* 2(1):22-33.

Talling, J.F. and Rzoska, J. (1967). The development of plankton in relation to hydrobiological regime in the Blue Nile *J. Ecol.* 55:636-662.

Tan, T.H. (1971). On occurrence and ecological adaptation of Zooplankton in Tanshi River Taiwan. *Rep. Inst. Fish. Biol. Taiwan,* 2(4):40-48.

Timms, B.V. (1970). Chemical and zooplankton studies of lantic habit in North, eastern New South Wales, *Aust. J. Mar. Freshwater, Res.,* 21:11-33.

Train, R.E. (1967). *In: quality criteria for water US Environmental protection Agency,* Washington.

Trivedi, R.K. and Goel, P.K. (1984). *Chemical and Biological Methods for water Pollution Studies.* Karad: Environmental Publications, 1-251.

Velz, C.J. (1947). Factory influencing self purification and their relation to pollution abatement (I) *Sewage work J.,* 19:629-644.

Venkateswarlu, T. and Jayanti, T.V. (1968). Hydro-biological studies of the river Sabarmati to evaluate water quality. *Hydrobiologia,* 33(3-4):442-448.

Venkateswarlu, T. (1969). An ecological study of the algae of the river Moosi Hyderabad (India) with special reference to water pollution I. Physico-chemical complexes. *Hydrobiologia,* 22(1):117-143.

Verma, M. (1964). Hydrobiological surveys of the Dalsagar tank, Seoni, Madhya Pradesh, India; with special reference to the effect of *Microcystis aeruguinosa* (Kutza) bloom on fish production and suggestions for the improvement of the tank fishery. *Ichthyologica,* 3:37-46.

Verma, S.R. and Shukla, G.R. (1969). Pollution in a perennial stream Khala, by the sugar factory effluent near Laksar (Dist. Saharanpur), U.P., *India. Env. Health,* 11:145-162.

Verma, S.R. *et al.* (1978). Pollutional studies of few rivers of western Uttar Pradesh with reference to biological indites. *Proc. Indian Acad. Sci.,* 87B:121-131.

Verma, S.R., Sharma, P., Tyagi, A., Rani, S., Gupta, A.K. and Dalela, R.C. (1984). Pollution and saprobic status of eastern Kalinadi, *Limnologica.,* 15(1):69-133.

Ward, R.W. and Seibert, H.C. (1963). Investigations into the limnology of Dow lake, Athens country. *Ohio. Jour. Sci.* 63(3):128-144.

Welch, P.S. (1948). *Limnological methods.* The Blakiston Co. Philadelphia, pp. 1-381.

Whitton, B.A. (1970). Toxicity of heavy metals to Fresh water algae, A review. *Phycos,* 2:116-125.

Whitton, B.A. (1983). *Ecology of European rivers.* Blackwell Scientific Publications. Oxford. 1-550.

Whitton, B.A. (1985). Biological monitoring of heavy metals in flowing waters. *Symp. Biomonitoring State of Environment*, 50-55.

Young, Y.C., Hannah, H.Y. and Mayhew, J.J. 1973. Nitrogen and phsphorous in a south of Guadalupe River, Texas with five main stream impoundments. *Hydrobiol.*, 43(3):419-441.

Zingde, M.D., Narvekar, P.V., Sharma, R.V. and Desai, B.N. 1980. Water Quality of the River Damanganga (Gujarat). *Ind. J. Mar. Sci.*, 9:94-99.

Zafar, Z.R. (1967). On the ecology of algae in certain fish ponds of Hyderabad, India III. Periodicity. *Hydrobiol.*, 30:96-112.

Index